Cell and Molecular Biology Lab Manual

David A. Thompson

Cell and Molecular Biology Lab Manual

David A. Thompson
Original artwork for cover: Cristina C. Thompson
Editing and review: Sharon L. Thompson

Publication date 2011
Copyright © 2011 David A. Thompson

Table of Contents

1. Laboratory reports 1
2. Safety 12
3. Statistics: correlations 13
4. Statistics: populations 20
5. Microscopy 27
6. Bacterial proliferation kinetics 35
7. Yeast proliferation kinetics 40
8. Neural networks 47
9. Protein concentration 68
10. SDS-PAGE 75
11. Western blot analysis 91
12. Reading a scientific paper 98
13. Restriction digest analysis 100
14. The PCR 109
15. Agarose gel electrophoresis of DNA 112
16. Physarum 122
A. Sample lab safety handout 127
B. Statistics software 134
C. Mass measurement 136
D. Microscopy 137
E. Sterile technique 149
F. Screenshots 152
G. Protein assays 153
H. Spectrophotometer use 157
I. Densitometry 162

List of Figures

1.1. Scatterplot of thymus mass data 7
1.2. Statistical summaries of thymus mass data 8
3.1. Scatterplot of height data 13
3.2. Histogram of height data 14
3.3. Scatterplot of height-sun data 18
3.4. Linear regression applied to height-sun data 19
4.1. Fixed-range 1000 µL pipettor 21
4.2. Fixed-range 40 µL pipettor 22
4.3. Thermometers with and without immersion lines 23
7.1. Schematic of the hemacytometer grid 45
8.1. A neural network 47
8.2. An optocouple circuit 47
8.3. A diagram of a neuron 48
8.4. The MyFirstNEURON start window 50
8.5. The MyFirstNeuron 'Parameters' window 52
8.6. The NEURON main menu window 53
8.7. The ArtCellGUI window 54
8.8. The NEURON spike plot window 57
10.1. A SDS-PAGE gel 78
10.2. A screwcap tube and cap 80
10.3. PAGE gel electrophoresis box 82
10.4. SDS-PAGE gel 83
10.5. Gel gaskets 85
11.1. Open transfer cassette with fiber pads and blotting paper 92
11.2. Schematic of assembly of Western transfer 94
13.1. Map of plasmid pTXB1/3 101
15.1. Visualization of apoptotic DNA 113

15.2. Disassembled minigel apparatus including wedges, comb, and tray 114
15.3. Tray in minigel apparatus 115
15.4. Tray with wedges in minigel apparatus 116
A.1. A safety shower and eyewash 129
D.1. A compound microscope 139
D.2. The microscope arm and stage controls 140
D.3. The microscope objective 141
D.4. The microscope stage 142
D.5. A slide on the microscope stage 143
D.6. The microscope light source 144
D.7. A simple condenser 145
H.1. The sample compartment and control panel of a spectrophotometer 158
H.2. A cuvet 159
H.3. The sample compartment of a spectrophotometer 160

List of Tables

1.1. Thymus masses at different mouse ages 6
9.1. Samples and standards 74
10.1. Example of a single row from a gel loading table 79
15.1. Example of planning for loading if one was evaluating sensitivity with an uncut plasmid 119

Chapter 1. Laboratory reports

Data collection

Before preparing a written scientific report, one typically has accumulated some data of interest and carefully recorded that data. Scientific data is recorded and stored in many different ways. If the records are likely to ever be associated with any legal proceeding, they should be kept meticulously in a bound notebook or using a computer-based record-keeping system. In such circumstances, after entering the data, the record can be verified with the signature and date of a witness or by using the verification functionality of a computer-based record-keeping system.

In an undergraduate lab, this type of record-keeping is overkill. Nevertheless, one should keep meticulous records. Your instructor may have specific expectations with respect to how you keep records. Such expectations might include the use of only ink in recording data, a specific manner in which any errors in recording are indicated and corrected, or the use of a specific medium for recordkeeping (e.g., a bound notebook).

Guidelines for preparing lab reports

Format guidelines

The report should be

- in blue or black ink, either clearly and neatly handwritten, typed, or printed from a computer

- left-justified, double-spaced, with one-inch margins on the left and right sides
- in a font which is normal weight 12 pt serif
- formatted for printing on 8.5 x 11 inch paper
- composed of numbered pages

Organization of the report

The report should be arranged with the following sections:

1. title, date of submission, author(s)
2. introduction
3. methods (experimental procedures)
4. results
5. discussion
6. references
7. figure legends
8. tables
9. figures
10. supplemental data

The sections of the report

Some of the texts listed in the 'selected references' section below provide excellent guidelines for writing a scientific paper. Familiarize yourself with these guidelines. Another approach to better understand the organization of a scientific paper and the content of each section is to read scientific articles.

Each section of the report should be clearly labeled. Prose should avoid a casual, conversational style and, instead, should be precise and concise.[1] Typically, this includes avoiding use of the first person.[2]

The **title section** should include the title (as short and informative as possible), the date of submission, and a list of authors (names of all authors and corresponding email addresses). Ensure that you exercise care in listing authors. For example, if the report describes a collaborative project, it may be appropriate to list several individuals as authors. However, it is likely that your instructor will expect you to clearly specify who actually wrote the text of the report.

The **introductory section** describes the subject of the study and the rationale/motivation underlying the study. Previous experiments leading to the current study are summarized (and the appropriate papers cited) in such a fashion that the reader is able to see how the train of logic progresses through the previous experiments to the question experimentally addressed in the present paper. Frequently this question is presented as an explicit hypothesis. In this course, you are not expected to complete a thorough literature search in order to write the introduction (unless specifically indicated); instead, a simple summary of

---

[1] An example of poor style: "Well, we conducted a study in our microbiology lab and started off by examining two restroom facilities..."
[2] For example, typically "...40 mL were heated..." is preferable to "...we heated 40 mL...".

the hypothesis of the study, the methodology used to test the hypothesis, and the conclusion drawn from the study will suffice.

The **methods section** describes the methodology used in the study. This section should be written in prose in the past tense and should be written in a sufficiently detailed fashion that the experiment could be reproduced by the reader.[3]

The **results section** describes the data obtained (the observations) from the present study. Typically, results are also presented in figures and/or tables. This section should include prose describing the data represented in the figures or tables as well as any other pertinent results. The prose should explicitly refer to figures and tables as appropriate.[4] Often, the results summarized in the "Results" section of a paper, are followed by one or several brief conclusions drawn from the results of each experiment. If a theoretical description (such as a mathematical equation) of the experimental system being studied is presented in the lecture or the lab handout, the results section is typically the appropriate place to apply it to your data. Frequent errors include omission of titles for figures or tables, omission of numbers for figures or tables, omission of figure legends, and omission of the prose component of the results section.

The **discussion section** summarizes the data and then proceeds to the author's interpretation of the data. Does the data support the hypotheses presented in

---

[3] Frequent errors include (1) composing this section as a list of numbered steps rather than writing it in prose and (2) using the imperative rather than the past tense.
[4] e.g., "Two bands of approximately 98 and 110 kDa were observed in lane 2 of the gel (Fig. 2)"

the introduction (or elsewhere...)? How will this study impact future studies?[5] Does the study have larger social or societal ramifications? Does the data allow the author to construct a model of how a particular biological process occurs?

Attach a copy of all raw data as well as any detailed calculations and formulas used during analysis of experimental data to the end of the report in the "supplemental data" section.

Tables and Figures

Each table or figure should

- be labeled with 'Table' or 'Figure' (if it's not a table, it's a figure...)

- be titled (e.g., "Table 2. Position versus velocity")

- be numbered (e.g., "Figure 3. Dependence of absorption on time")

- be accompanied by a descriptive legend

Figures should have descriptive axis labels (including units of measure, when appropriate).

Presenting data

Whether one continues in science or enters another field, it is important to understand how to interpret and present data in the context of "graphs" or "charts".

---

[5]In other words, what questions do the data suggest should be asked next? How would those questions be addressed experimentally in future studies?

Interpreting raw data typically requires a substantial investment of effort. One of the author's "jobs" is to reduce the amount of effort the reader must invest in order to visualize data. When done well, the presentation enables the reader to comprehend trends in data and aids the reader in evaluating the conclusions drawn with respect to the hypothesis/es presented by the author.

Table 1.1. Thymus masses at different mouse ages

week	mass (g)					
8	0.03	0.05	0.07	0.09	0.06	
12	0.03	0.05	0.07	0.08	0.12	0.19
16	0.05	0.07	0.07	0.09	0.15	
20	0.05	0.07	0.08	0.10	0.20	

For example, consider a (hypothetical) study of thymus mass versus mouse age. While thymus mass tends to rise with age, this trend isn't immediately apparent from inspection of the table of raw data. Neither does a scatterplot of the data (Figure 1.1, "Scatterplot of thymus mass data") forcefully drive this point home. If it is desirable to emphasize the trend, the point might be driven home more forcefully by using a figure with statistical summaries (Figure 1.2, "Statistical summaries of thymus mass data").

Figure 1.1. Scatterplot of thymus mass data

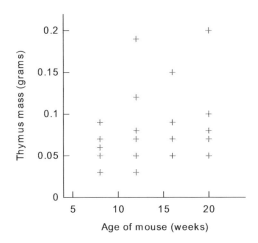

Mice were sacrificed at the indicated ages. The wet weight of the thymus of each sacrificed mouse (solid cross) was determined immediately after euthanization of the mouse.

Figure 1.2. Statistical summaries of thymus mass data

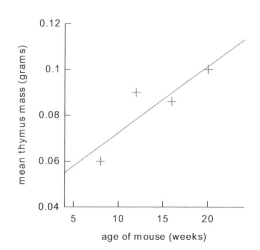

Mice were sacrificed at the indicated ages. The wet weight of the thymus of each sacrificed mouse was determined immediately after euthanization of the mouse. A least-squares linear regression was performed (solid line) using averages at each time point.

References

When material from another source is used in the report, the source must be acknowledged. This is done through (1) an in-text citation and (2) the inclusion of a reference list (typically near the end of the document). Use the APA citation and reference list styles (see http://www.apastyle.org/) when preparing a report.

Acceptable references and sources are articles in peer-reviewed journals or peer-reviewed texts. The primary reference should be used whenever possible. Other references and sources are generally not acceptable.[6]

---

[6] For example, textbooks are generally not acceptable references. Neither are most "web pages" (e.g., wikipedia articles) acceptable references.

If an excerpt from another work is used verbatim, it must be clearly formatted as such. Short quotes can be indicated by surrounding the quoted material with double-quote (") characters. Longer quotes should be formatted as indented blockquotes. Both short and long quotes should be followed by an appropriate citation.

Additional resources

Selected references for scientific writing

The craft of scientific writing. 1996. By Michael Alley

Successful scientific writing: a step-by-step guide for the biological and... 2007. By Janice R. Matthews and Robert W. Matthews

Writing scientific research articles: strategy and steps. 2009. By Margaret Cargill and Patrick O'Connor

How to write and publish a scientific paper. 2008. By Robert A. Day

How to write and illustrate scientific papers. 2008. By Björn Gustavii

A coursebook on scientific and professional writing for speech-language... 2008. By Mahabalagiri N. Hegde

Scientific Writing = Thinking in Words. 2011. By David Lindsay

Word-processing software

LibreOffice [http://www.libreoffice.org/]

OpenOffice [http://www.openoffice.org]

AbiWord [http://www.abisource.com]

KOffice [http://koffice.org]

Plotting software

gnumeric [http://projects.gnome.org/gnumeric/]

gnuplot [http://www.gnuplot.info]

LibreOffice [http://www.libreoffice.org/]

OpenOffice [http://www.openoffice.org]

KOffice [http://koffice.org]

Other resources

Create A Graph [http://nces.ed.gov/nceskids/createAgraph/]

- a NCES site which facilitates experimenting with graph generation

How to Lie With Charts (2000)

- a text by Gerald Everett Jones

Edward R. Tufte is a well-known data visualization guru. Who would have thought one could say so much about charts and graphs? An article for the 'layperson' describing Tufte's work is here [http://www.salon.com/march97/tufte970310.html]. Tufte's works include:

- The Visual Display of Quantitative Information (2001)

- Envisioning Information (1990)

Laboratory reports

- Beautiful Evidence (2006)

- Visual Explanations: Images and Quantities, Evidence and Narrative (1997)

Chapter 2. Safety

Before beginning any work in the laboratory, the course instructor should discuss the laboratory rules[1] and any specific safety issues associated with the laboratory component of the course. Some lab procedures may have specific safety considerations; these will be noted in the laboratory manual and/or described by the course instructor prior to the lab.

---

[1] Appendix A, *Sample lab safety handout* contains an example of some guidelines which might be enforced in the laboratory component of a university course.

Chapter 3. Statistics: correlations

Statistics

Visual and quantitative descriptions of a population

A data set can be described using the visual aid of a "chart" or "plot" in a number of ways. Consider a simple set of data corresponding to the heights of individual plants in a population of dandelions: 10.0, 9.5, 9.4, 11.1, 4.6, 10.4, 12.1, 11.4, 11.2, 10.7, 9.9, and 15.1 cm. The same information can be conveyed in several ways. Inspection of the "raw data" above doesn't immediately give a sense of whether the heights are clustered about a given region nor whether "outliers" exist. In contrast, clustering and outliers are readily visualized with a scatterplot Figure 3.1, "Scatterplot of height data" or with a histogram Figure 3.2, "Histogram of height data".

Figure 3.1. Scatterplot of height data

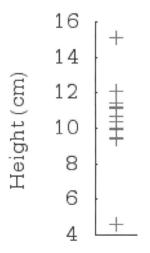

Figure 3.2. Histogram of height data

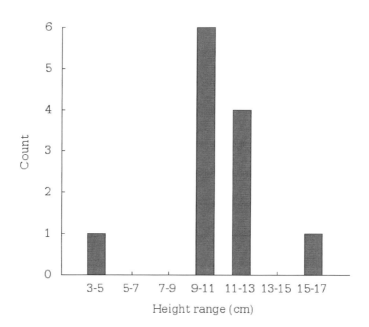

Along with a visual tool to aid in evaluating the data points in a population, it may be desirable to calculate values, *statistics*, which serve as useful descriptors of that data set. Two qualities often of interest are a measure of the 'central tendency' (intuitively, the "middle" value) and a measure of the extent of variation within a population.

The **mean**, although susceptible to the influence of outliers, is often used as a measure of central tendency. The **standard deviation** is often used as a measure of the variation within a population. Both values can be readily calculated using calculators with statistical functionality or with most spreadsheet programs.

Exercise:

- calculate the mean of the dandelion heights

Statistics: correlations

- calculate the standard deviation of the dandelion heights

- calculate the mean and standard deviation for a second population of dandelions with heights 10.0, 9.5, 9.4, 11.1, 8.6, 10.4, 12.1, 11.4, 11.2, 10.7, 9.9, and 11.1 cm

- compare the mean and standard deviation values for each population, explaining any differences

Influence of sample size on estimate confidence

Consider a relatively simple sampling situation, the binomial distribution. This distribution describes a series of 'trials' (independent and identical) where each trial has two possible outcomes. An oft-utilized example is the series of coin tosses. For more biological flavor, consider an experiment in which 1000 cells (out of a much larger population) are counted under the microscope for apoptotic character and 50 are found to be apoptotic. We can calculate statistics which describe:

1. an estimate of the actual percentage of apoptotic cells in the population

2. a measure of our confidence in that estimate

p = estimate of probability of event = 50/1000

How much confidence do we have in this value? Note that a count of 5/100 would yield the same value for p. Intuitively, we recognize that we have more confidence in the estimate derived from the 1000-cell count than in the estimate derived from a 100-cell count. How does one quantify this? A **confidence interval** is a range that "traps" or covers a population parameter with a specified probability. For example, the 95% confidence interval for p(50/1000) is 0.0373 to 0.0654. What does this value represent? If the sampling process is repeated, then, given our initial observations, we expect that 95/100 times, we would obtain a value between 0.0373 and 0.0654.

Exercise:

- what is the 95% CI for p(5/100)? Compare it to the confidence interval for p(50/1000)

"Take-home message": increasing the size of the sample increases "confidence" in the estimate obtained from that data set.

Making comparisons

So far, we have considered single populations in isolation. Furthermore, for each population we have only considered a single variable.

comparing means of two populations

How does one quantitatively compare the strength of similarity or difference between the means of two populations? Such a comparison must take into account not only the means themselves but also the variation present in each population. Things can get a bit murky with the variety of possible permutations of knowns and unknowns (are the standard deviations known?)

Statistics: correlations

and whether or not the members of the populations are paired.

For a situation where means and standard deviations are unknown, and population members are not paired, a "two-sample t statistic" may be used. In this case, t statistic for the difference in means, $mean_A - mean_B$, is given by

$$t = \frac{mean_A - mean_B}{\sqrt{\frac{stdev_A^2}{n_A^2} + \frac{stdev_B^2}{n_B^2}}}$$

where

$mean_A$ is the mean of data set A

$stdev_A$ is the estimate of the standard deviation of data set A

n_A is the number of measurements in data set A

The t statistic may be evaluated based on the number of degrees of freedom ($n_A + n_B - 2$) using either a t statistic table or a program which calculates the corresponding confidence level. When using a table, locate the column or row corresponding to the calculated degrees of freedom value. Then locate the corresponding t statistic value and, from that, the corresponding confidence level at which the hypothesis (that the means are equal) can be rejected.

correlations between two variables

A correlation between variables can be visualized using a tool such as a scatterplot. In a relatively simple case, both variables are continuous numeric variables, and the correlation can be modeled with linear regression. In such a case, the "strength" of such a correlation is represented by a **regression coefficient**. Other modeling approaches may be required if the analysis considers more than two variables or if any of the variables are not continuous numeric variables.

Figure 3.3. Scatterplot of height-sun data

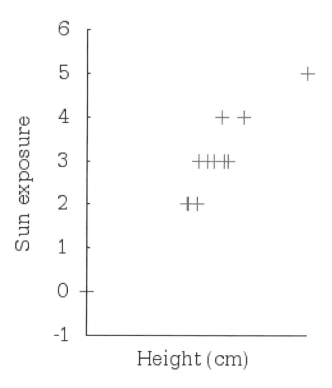

Figure 3.4. Linear regression applied to height-sun data

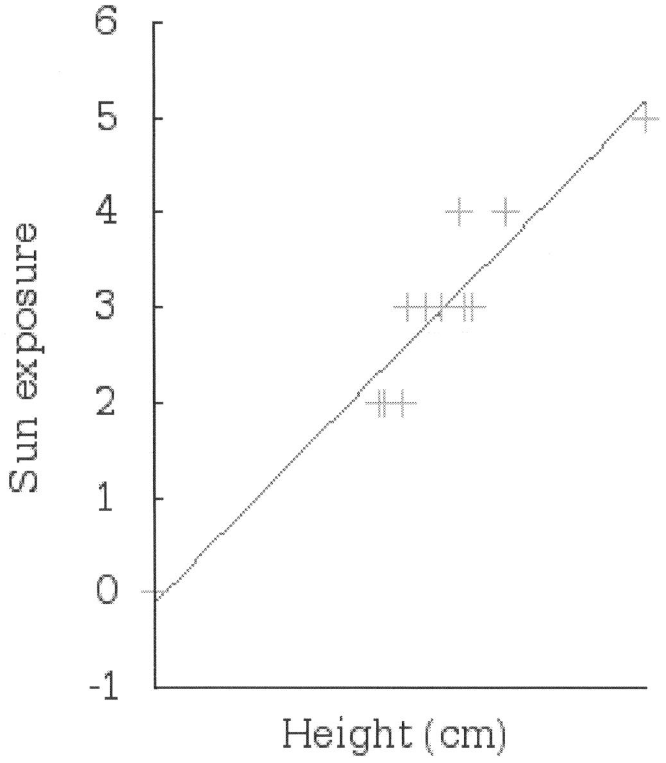

Resources

Various statistics software packages available online are described in Appendix B, *Statistics software*.

Woolf P, A Keating, C Burge, and M Yaffe, 2004, Statistics and Probability Primer for Computational Biologists, Massachusetts Institute of Technology, http://ocw.mit.edu/NR/rdonlyres/Biology/7-89Fall-2007/Readings/

Chapter 4. Statistics: populations

Obtaining, recording, and presenting data

Background

In this experiment, you will evaluate tools that you will use during the course. In conducting an experiment, the quality of data obtained depends on the quality of the measurement. The quality of a measurement is related both to the technical skill of the individual taking the measurement and to the quality and calibration of the instrument employed to make the measurement.

In science, the accuracy and precision of measurements can be a critical determinant of the strength of the conclusion which can be drawn regarding a hypothesis. It is equally important to have the ability to present data in a manner that communicates clearly the "take-home message". In both academia and business, such presentations are regularly made to one's peers, one's employer, or funding source(s). It becomes important not only to have quality data, but to have the ability to convey the message behind the data in a convincing and clear fashion.

Procedure

Evaluate calibration of pipettors

1. Obtain a 1000 µL fixed-volume pipettor and tips for the pipettor.

Statistics: populations

Figure 4.1. Fixed-range 1000 µL pipettor

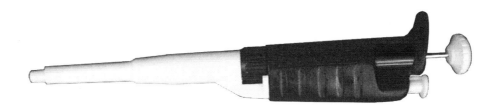

2. If the pipettor has not already been labeled (a piece of tape or sticker with an "identity" written on it), place a small piece of masking tape on the side of the pipettor and label it with a unique name. Once the pipettor is labeled, record the identity of the pipettor.

3. Tare a piece of weigh paper or a weigh boat and add 1000 µL water to the paper; record the measurement.

4. Tare the same piece of paper, add 1000 µL and record the measurement; repeat the process a third time.

5. Ensure that for the subsequent mass measurements, you are using a balance with at least ± 0.001 g resolution.

6. Obtain a fixed-range 40 µL range pipettor and tips for the pipettor.

Figure 4.2. Fixed-range 40 µL pipettor

7. Record the identity and volume of the pipettor.

8. Tare a piece of weigh paper or a weigh boat and add 200 µL water to the paper; record your measurement.

9. Repeat the measurement twice (as above).

10. Obtain a 10 µL fixed-volume pipettor and tips for the pipettor

11. Tare a piece of weigh paper or a weigh boat and add 10 µL of water to the paper; record the measurement.

12. Repeat the measurement twice (as above).

13. Repeat the above series of 10 µL measurements (three measurements) using DMSO.

Evaluate thermometer calibration

Figure 4.3. Thermometers with and without immersion lines

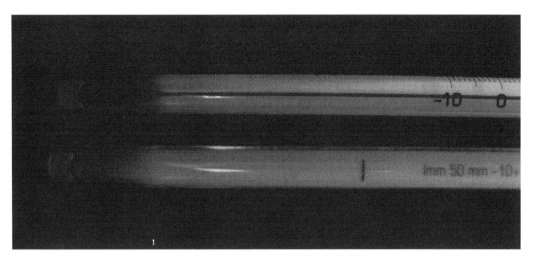

1. Obtain a thermometer with at least a 0-100°C scale.

2. If the thermometer is not labeled, wrap a piece of tape around the extreme (non-measuring) tip of the thermometer. Label the thermometer with a unique label (e.g., your initials and today's date)

3. Dispense a volume of water sufficient for the thermometer you're using (look for the 'immersion line' on the thermometer (Figure 4.3, "Thermometers with and without immersion lines")) into a 250 mL beaker (you may want to use a 125 or 250 mL Erlenmeyer flask depending on the characteristics of the thermometer you are using). Heat the water to boiling (e.g., using a hot plate or a microwave).

4. Measure the temperature three times.[1]

    • let the thermometer cool to room temperature before and between each measurement[2]

Statistics: populations

- you may wish to use tongs or insulated gloves to hold the thermometer when working with hot liquids

5. Pour the hot water into the sink.

6. Fill the empty beaker approximately one-third to one-half full (about 100 mL water) with room-temperature distilled water.

7. Add ice to the water to saturation (i.e., the ice should not be rapidly melting); swirl the beaker for 30 s.

8. Measure the temperature three times; let the thermometer warm to room temperature between each measurement.

The lab report

Make sure to present the data from the experiment (including, but not limited to):

- raw data (should be presented in tables)

- statistical summaries (e.g., the mean and standard deviation of a series of measurements of the mass of 1 µL aliquots of water) and measures (you may find a table is a convenient means of presenting this data)

Remember to evaluate the data in the discussion. For example,

- characterize the calibration of the pipettors (adequate? inadequate? for what?)

- characterize the calibration of thermometers (adequate? inadequate? for what?)

Discuss possible sources of error in measurements. e.g., were some of the measurements more highly variable than others? (e.g., larger standard deviations as a percentage of the mean?) -- did some measurements not match anticipated values?

Define <u>accuracy</u> and <u>precision</u>; characterize each instrument you evaluated in terms of its accuracy and precision.

Additional resources

Evaluating mass

Appendix C, *Mass measurement*

Statistics

A standard undergraduate statistics text typically has tables in the back representing the cumulative standard normal and cumulative binomial distributions. Many modern calculators that include most standard statistical functions can be purchased relatively inexpensively. Common computer programs such as the spreadsheet "Excel" also have most standard statistical functions. There are also quite a few online tools for calculating statistics (see Appendix B, *Statistics software*).

Chemical constants

The Merck Index, the CRC Handbook, and a variety of chemical catalogs (e.g., the Sigma-Aldrich catalog and the ICN catalog) contain physical constants (e.g., density, melting point, boiling point) for many compounds.

Chapter 5. Microscopy

See Appendix D, *Microscopy* for an introduction to light microscopy.

The laboratory report

1. The laboratory report should follow the format described in the first laboratory session.

2. Make sure to present data from the experiment (see below)

3. Evaluate the quality of the observations made

    - were there deficiencies in instrumentation? in sample preparation technique or in other technical aspects of the experiment?

Prelab

1. Before the laboratory session, browse the guide outlining steps in using the specific microscopes in your laboratory (if your instructor made such a guide available...).

2. Answer the following questions on a separate sheet of paper and turn it in at the start of the laboratory session.

    Why would it be advantageous for a microscope to have a filter which allows only blue light to pass through the sample?

    What is the purpose of immersion oil?

Where does the optical path of the light microscope begin?

What is the function of the diaphragm?

What is the function of the condenser?

3. Read the guidelines in Appendix D, *Microscopy* before using the microscope

Refamiliarizing yourself with the compound microscope

Read 'Familiarizing yourself with the compound microscope' (in Appendix D, *Microscopy*) through the 'objectives' subsection and then...

- **obtain a dry mount of a highly stained sample (i.e., you can visualize the sample on the slide with your unaided eye) with good contrast for focusing purposes during the subsequent steps**

  - **place the lowest power objective in the light path and move the stage away from the objective at least 1 cm**

the stage

Read 'Familiarizing yourself with the compound microscope' (in Appendix D, *Microscopy*) through the 'the stage' subsection and then...

- **place the slide on the stage** [1]

the light source

Read 'Familiarizing yourself with the compound microscope' (in Appendix D, *Microscopy*) through the 'the light source' subsection and then **switch on the light source and adjust to a low intensity**.

the condenser

Read 'Familiarizing yourself with the compound microscope' (in Appendix D, *Microscopy*) through the 'the condenser' subsection and then follow the procedure for adjusting the condenser.

Read 'Familiarizing yourself with the compound microscope' (in Appendix D, *Microscopy*) through the 'the interpupillary distance' subsection and then follow the procedure for adjusting the interpupillary distance.

Ocular micrometer calibration

- obtain a stage micrometer (a glass slide with etchings)

- note that the standard stage micrometer is labeled "0.01 mm"; the etched bar on the slide is 1 mm long; the bar is divided into 100 subdivisions

- place the micrometer on the stage and focus on the micrometer gradations

- use the micrometer to determine the length corresponding to the eyepiece gradations in the eyepiece scale at 4X and 10X

Considerations when examining cells via microscopy

Microscopy

In the following parts of this exercise, you will use microscopy to observe fixed and/or live cells. Carefully record your observations. Examine the cells at both low and high power. Draw a sketch of a representative cell or set of cells. Note the following features:

- How big is the cell?

- Is the cell relatively flat or does it appear to have significant depth?

- What shape is the cell?

- Is there a cell wall?

- Which organelles are visible?

- Is the cell colorless or colored?

- If multiple cells are present in the sample, are all the cells the same type? If not, describe the qualities which distinguish the cell types.

Evaluating dry mounts

- evaluate one dry mount of smooth muscle cells, including an estimate of the size (length and width) of a single smooth muscle cell

- obtain dry mounts of two different types of bacteria

- using the oil immersion lens may be necessary for visualizing the bacteria as clearly as possible; to use the oil immersion lens:

  - place a drop of immersion oil carefully on top of the cover slip

Microscopy

- place the slide on the stage

- rotate the oil lens until it touches the oil and the oil spreads across the lens surface, contacting both the cover slip and the objective lens

- carefully bring the sample into focus

- clean the lens and the slide (use lens paper and isopropanol or ethanol) when finished

Evaluating wet mounts

- evaluate a wet mount of cheek or tongue squamous epithelial cells:

  - place slide on a flat surface

  - scrape the inside of the cheek (fairly forcefully, but no need to draw blood...) with the flat end of a toothpick

  - transfer the material on the toothpick to the slide

  - hold the coverslip by its sides and lay its bottom edge on the slide close to the specimen

  - slowly lower the coverslip so that it spreads the liquid out. If you get air bubbles (looking like little black doughnuts), gently press on the coverslip to move them to the edge

  - using the data from the optical micrometer, estimate the length of a typical squamous epithelial cell from the cheek

Microscopy

- under low and high power attempt to sketch and label the nucleus, the darkly staining nucleolus, the nuclear membrane, the cell membrane, and the cytoplasm

- look for bacteria

  - do you see any (e.g., adhering to the cheek cells)? if so, can you distinguish the types observed?

- try staining the specimen:

  - Lugol's iodine, methylene blue, or crystal violet[2] may be added to specimens in order to increase contrast

  - stain can be added later after first viewing the specimen without the stain:

    - Add a drop of the stain along a coverslip edge

    - Placing a Kimwipe along the opposite edge of the coverslip will help draw the stain under the coverslip.

Fluorescence microscopy of epithelial cells

Evaluate a wet mount of cheek or tongue squamous epithelial cells by fluorescence microscopy using Hoechst 33258, a DNA-specific stain:

- harvest cells as above but don't add cover slip

- add a drop of 1 µg/mL Hoechst 33258 solution (it contains a few grains of Carnation dry milk as well) on the slide on the region where the sample was placed (note that Hoechst 33258 is membrane-permeable)

- hold a coverslip by its sides and lay its bottom edge on the slide close to the specimen

- slowly lower the coverslip so that it spreads the liquid out. If you get air bubbles (looking like little black doughnuts), gently press on the coverslip to move them to the edge

- incubate 5 min in the dark

- note: if the sample exhibits photobleaching, add a 1/4 drop of 0.1M N-propyl gallate (in 80% glycerol/20% water)

- evaluate via fluorescence microscopy - in particular, look for:

    - bacteria: do you see any bacterial cells (e.g., adhering to the cheek cells)? if so, can you distinguish the types observed?

    - cells at different phases of the cell cycle (are there any cells that appear to be mitotic? apoptotic?)

Evaluating a wet mount of plant cells

Evaluate a wet mount of onion cells:

- Cut a red onion into sections

- Remove an onion layer (a "leaf") (these are the large (appr. 3 mm thick) concentric rings of the onion)

- Grab the two ends of the leaf and bend them until the leaf snaps

- Peel off a thin piece of the inner epidermis

- Place the thin piece in a drop of water on a microscope slide and place a coverslip over the sample

- Place a small drop of methylene blue or Lugol's iodine at the edge of the coverslip and draw the stain through the sample by wicking, using a piece of tissue paper (or Kim-Wipe) at the other side of the coverslip to draw the fluid across/under the coverslip

- After staining for approximately 5 min, focus to visualize the stained cytoplasm surrounding the vacuole of a cell

- Attempt to identify the nucleus of a cell.

- Sketch and label a stained cell in the space in the text

- Try staining with a different stain such as methylene blue

Cleanup

All slides should be disposed of in the appropriate containers:

- biohazard material: anything which has come in contact with a bodily fluid

- non-biohazard slides and cover slips can be disposed of in the broken glass box

Chapter 6. Bacterial proliferation kinetics

Introduction

The capacity of a bacterium to proliferate varies across species as well as with environmental conditions. Bacterial proliferation kinetics can have an enormous impact in a variety of areas (e.g., the clinical course of an infection, the production of a recombinant protein, or on the spoilage of foods). The typical *E. coli* proliferation curve comprises a lag phase of approximately 1.5 h, a period of exponential proliferation, a deceleration phase, a stationary phase, followed by a decline (or "autolytic") phase. One of the more straightforward ways to monitor bacterial proliferation in liquid culture is by use of spectrophotometry, in our case the measurement is essentially of turbidity. When light is directed onto a suspension of bacterial cells, light is scattered when it encounters a bacterial cell, resulting in less light passing through the suspension as the number of cells rises. A complete analysis of the growth curve requires a correlation of the optical density (OD) with the actual cell number. Different *E. coli* strains and genotypes may vary substantially in the length of the various phases of the proliferation curve.

Spectrophotometry

An instrument that measures the amount of absorption (A), often referred to as optical density (O.D.), of a solution is called a spectrophotometer. This spectrophotometer passes a beam of light of a specific wavelength through the sample (contained in a cell called a cuvet). The light that passes through the cell is quantified by a photodetector on the far side of the cell. The greater the concentration of the absorbing component of a sample, the more light is absorbed by the sample. The relationship between concentration and the absorption of light by the sample is represented by the Lambert-Beer law:

Transmittance = $I/I_0 = e^{-kcl}$

where I_0 is the intensity of the incident light, I is the intensity of the light reaching the photodetector after passing through a distance l (usually 1.0 cm) of solution of concentration c having a characteristic absorption coefficient k, a constant.

Absorbance (also termed the optical density) is typically used to measure the extent of light absorption by a sample. The relationship between absorbance and transmittance T is given in the following formula:

A = log (1/T) = log (I_0/I) = kcl

Note the linear relationship between absorbance and concentration.

Exponential proliferation

When proliferation is not limited by the composition of the medium or other factors, cells are said to be in "exponential phase" or "logarithmic" proliferation: in each successive time period, the size of the population increases by a constant multiplicative factor. Represented mathematically,

Bacterial proliferation kinetics

$dN/dT = k_p N$

where N is the number of cells present at time t and k_p is a proliferation constant[1] that depends on cell type and environmental conditions.

One can rearrange this formula as

$dN/N = k_p \, dt$

Integrating between (N_o, t_o) and (N, t):

$\ln(N/N_o) = k_p(t - t_0)$

The "population doubling time" is the value of t for which $N/N_o = 2$. Moreover, a plot of $\ln(N/N_o)$ as a function of t will be a straight line whose slope is k. "k" can be estimated for a set of data points using a standard graphing program equipped with regression analysis tools. The doubling time estimated from that set of data points is then relatively straightforward to calculate (that calculation is left to the reader).

Procedure

1. Turn the spectrophotometer on

2. Prepare a 1:20[2] (volume:volume) dilution of an "overnight culture"[3] of bacteria:

- remove the foil or cap from the flask containing the overnight culture; flame the upper neck of the flask to reduce the probability of contaminants entering the flask

---

[1] k_p is used to avoid confusion between this constant and the absorption coefficient k in the Beer-Lambert law

Bacterial proliferation kinetics

- withdraw 1 mL; transfer that volume to a 50 mL Erlenmeyer flask containing 19 mL of Luria-Bertani broth (LB) (flame the flask neck after removing foil from this container)

- mix thoroughly; withdraw 1000 µL of the 20-fold dilution and place in a cuvet - cover the cuvet opening with parafilm

- note the time; replace the foil or cap[4] on the 20-fold dilution culture and place the culture in the shaking incubator at 37°C

- evaluate transmittance at 550 nm or A_{550} (shorthand for absorbance at 550 nm) of the sample you just took (make sure you autozero on a reference, e.g., water, first)

1. At 0 min, 45 min, 1.5 h, 3 h, 6 h, and at 9/15 h[5] (time with respect to initiation of the bacterial culture which you carried out in step 1), measure the transmittance or absorbance of the culture (this is described in step 1 for the 0 min time point):

- remove foil from the mouth of the culture flask

- flame upper neck of flask (note: at later time points, this step is actually not essential since any contamination would not substantially interfere with the measurement)

- withdraw 800 µL of the bacterial culture and transfer it to a 1 mL cuvet

- evaluate A_{550}

---

[4] ensure that the bacteria can 'breathe' -- we're evaluating aerobic proliferation

Bacterial proliferation kinetics

- immerse the cuvet in 10% bleach for at least 30 min; rinse with water and dry (invert on a paper towel) so there are no water marks

Laboratory report

Ensure you evaluate your data; for example:

- calculate k_p (show formulas and calculations)

- what is the doubling time of the bacterial strain under analysis? show your formulas and calculations

- what is the length (with respect to time) of each phase of the bacterial proliferation curve for this bacterial strain?

Discuss the data -- for example, how does the observed doubling time compare with the typical doubling time of mammalian cells in culture?

Address the following question:

Mammalian cells are generally cultured in nutrient-rich liquid media. Why must mammalian cell culture be conducted under highly sterile conditions for most experimental purposes?

References

"Exercises in cell biology for the undergraduate laboratory" (Ed. Ledbetter, M., [http://ascb.org/pubs/exercises.html], 1992)

Chapter 7. Yeast proliferation kinetics

Introduction

S. cerevisiae is widely used both in food applications such as brewing and baking and in scientific applications. These simple eukaryotes can be used to maintain genetic vectors and to study eukaryotic genetics.

spectrophotometry

An instrument that measures the amount of absorption (A), often referred to as optical density (O.D.), of a solution is called a spectrophotometer. This spectrophotometer passes a beam of light of a specific wavelength through the sample (liquid samples are contained in a cell called a cuvet). The light that passes through the cell is quantified by a photodetector on the far side of the cell. The greater the concentration of the absorbing component of a sample, the more light is absorbed by the sample. The number referring to this phenomenon is the "absorbance" (also termed the optical density). The relationship between concentration and optical density is represented by the Lambert-Beer law:

Transmittance = $I/I_0 = e^{-kcl}$

where I_0 is the intensity of the incident light, I is the intensity of the light reaching the photodetector after passing through a distance l (usually 1.0 cm) of solution of concentration c having a characteristic absorption coefficient k, a constant. Optical density is usually measured in absorbance units because a linear relationship exists between absorbance and concentration. The relationship between absorbance and transmittance is given in the following formula:

$A = \log(1/T) = \log(I_o/I) = kcl$

Exponential proliferation

When proliferation is not limited by the composition of the medium or other factors, cells are said to be in "exponential phase" or "logarithmic" proliferation: in each successive time period, the size of the population increases by a constant multiplicative factor. Represented mathematically,

$dN/dT = k_p N$

where N is the number of cells present at time t and k_p is a proliferation constant[1] that depends on cell type and environmental conditions.

One can rearrange this formula as

$dN/N = k_p\, dt$

Integrating between (N_o, t_o) and (N, t):

$\ln(N/N_o) = k_p(t-t_0)$

The "population doubling time" is the value of t for which $N/N_o = 2$. Moreover, a plot of $\ln(N/N_o)$ as a function of t will be a straight line whose slope is k. "k" can be estimated for a set of data points using a standard graphing program equipped with regression

---

[1] k_p is used to avoid confusion between this constant and the absorption coefficient k in the Beer-Lambert law

analysis tools. The doubling time estimated from that set of data points is then relatively straightforward to calculate (that calculation is left to the reader).

Procedure

1. Prepare (approximately) a 1:60 dilution of an "overnight culture"[2] of *S. cerevisiae*:

    - prepare a sterile 50 mL Erlenmeyer flask containing 15 mL YPD

    - remove the lid or foil "cap" from the flask containing the "overnight culture"; flame the upper neck of the flask

    - withdraw 200 µL of the overnight culture and transfer the 200 µL to the 15 mL YPD in the 50 mL flask

    - place the 50 mL flask in the 30°C shaking water bath

2. For afternoon labs, measure the absorbance at 550 nm of a 3:2 dilution (volume:volume) in water of the culture at the following time points: 0, 1, 2, 3, 5.5, 16-18, and 24-26 h (time with respect to initiation of the diluted culture)

    - remove the foil or threaded lid from the mouth of the culture flask

    - transfer 600 µL of the culture to a cuvet

    - use the threaded lid or foil to cover the mouth of the culture flask, ensuring that the culture

Yeast proliferation kinetics

flask is *not* sealed tightly (unless the goal is to collect data regarding proliferation under anaerobic conditions)

- add 400 µL water to the cuvet; mix the cuvet contents thoroughly (using a clean glove surface or Parafilm to cap the cuvet)

- evaluate A_{550} using the spectrophotometer

- return the culture flask to the 30°C incubator

3. At the initial, fourth, and seventh time points, evaluate the yeast concentration in your culture using the hemocytometer:[3]

    - set the microscope on the low power objective

    - *carefully* remove the hemocytometer and coverslip from the container

    - place the coverslip over the hemocytometer chamber

    - remove a 50 µL sample and slowly let it flow into the hemocytometer chamber by capillary action until the chamber is full (you may not need the entire 50 µL)

    - find the grid pattern at 40X total magnification (see Figure 7.1, "Schematic of the hemacytometer grid")

    - "zoom in" on the upper left-hand corner square with 100X total magnification (you should see a 4x4 grid)

Yeast proliferation kinetics

- count the cells in that 4x4 grid

- if you count less than 10 or more than 100 cells per square, prepare a new dilution and repeat the above steps

- calculate the number of cells/mL: (average count per square) x 10^4 x (dilution factor)

- thoroughly rinse the coverslip and the hemocytometer and dry them gently with a paper towel

4. At the initial, fourth, and seventh time points, examine wet mounts of the culture via microscopy, preferably using phase contrast optics with a 10X objective or 40X objective. The same sample used for counting with the hemocytometer can be used again to collect this data. Count at least 50 cells and categorize each as "No bud", "Small bud", "Large bud", or "Mitotic".

5. Optional: Evaluate the yeast concentration by plating serial dilutions of the yeast culture

    - bring 2 YPD-agar dishes to room temperature

    - prepare two sterile microcentrifuge tubes, each containing 198 µL of sterile YPD

    - transfer 2 µL from spectrophotometer cuvet containing initial sample to the first microcentrifuge tube; mix thoroughly (vortex); withdraw 2 µL from the first tube and transfer to second microcentrifuge tube

Yeast proliferation kinetics

- transfer 100 µL from the cuvet to the first YPD-agar dish

- transfer 100 µL from each Eppendorf tube into the 2nd and third YPD-agar dishes (label the dishes with distinct identifiers!)

- disperse cells on the dish: flame a spreader made from a Pasteur pipette; cool the spreader by touching it to the agar; spread cells by back-and-forth motion followed by rotating motion

- place Petri dishes in an incubator (30°C)

6. Don't forget to turn off the spectrophotometer after data has been collected for the last time point.

Figure 7.1. Schematic of the hemacytometer grid

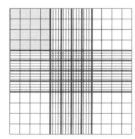

Laboratory report

Ensure that you evaluate your data (including, but not limited to):

- what is the doubling time of the yeast strain under analysis (show your formulas and calculations)? How does this compare with the typical doubling time of mammalian cells in culture? bacterial cells in culture?

Yeast proliferation kinetics

- if the absence of a bud corresponds to a cell in G1 phase, a small bud corresponds to a cell in S phase, a larger bud corresponds to a cell in G2/M phase, what is the cell cycle distribution of the population you evaluated by light microscopy? is this consistent with the time at which it was sampled?

- what is the length (with respect to time) of each phase of the yeast growth curve for this yeast strain under near-optimal conditions?

- compare your cell counts, spectrophotometry data, and cell cycle distribution data; do they seem consistent with each other?

- estimate the length of each phase of the yeast cell cycle

Other questions:

- why was 550 nm used to measure the turbidity of the culture?

- what do you think the historical/original purpose of the hemocytometer was?

Chapter 8. Neural networks

Figure 8.1. A neural network

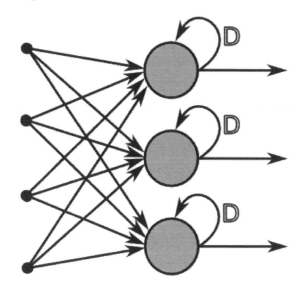

Figure 8.2. An optocouple circuit

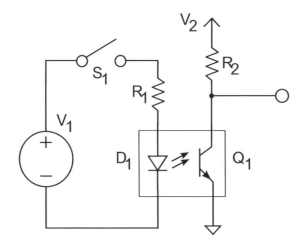

Figure 8.3. A diagram of a neuron

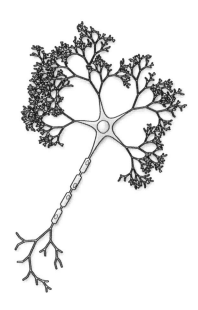

Introduction

The human brain contains approximately 10^{11} neurons. Huttenlocher (Huttenlocher, 1979) estimated a synaptic density of 11.05 x 10^8 synapses/mm^3 in the adult human, an estimate implying thousands of synapses per neuron. These synaptic connections are very specific: the neuron has many inputs but only one output (the axon). Apart from being composed of cells, this high degree of connectivity is one of the remarkable aspects of the brain differentiating it from a typical digital computer.

To begin to understand how the brain functions, one must first understand how neurons communicate with each other. A neuron sends and receives signals either electrically, via gap junctions, or chemically, via neurotransmitters and their corresponding receptors located at the site of signaling, the synapse. Signaling at the synapse is complex: the postsynaptic neuron receives signals from other neuron(s) and, if the "accumulation" of those signals exceeds a given threshold, the neuron fires.

Biological neural networks can be simulated on computers. Typically the approach involves defining single units, "nodes", which are the analogs of neurons. Such a network is typically meant to be a simulation of a central nervous system, typically with the goal of replicating CNS activities such as memory, pattern recognition, decision making, etc. Technological limitations restrict present-day simulations to much smaller sizes than the actual network of the human brain.

However, even such smaller networks may be used to improve the performance of critical real-word tasks. For example, El-Solh *et al.* concluded that an artificial neural network can identify patients with active pulmonary TB more accurately than physicians' clinical assessments[1] and Burke *et al.*[2] concluded that an artificial neural network can provide a useful "second opinion" in the diagnosis of lung disease.

Exercise: simulating an action potential

The lab report for this lab should simply consist of the answers to the questions posed below along with printouts where indicated.

---

[1] El-Solh AA, C Hsiao, S Goodnough, J Serghani, and BJB Grant, 1999, Predicting active pulmonary tuberculosis using an artificial neural network, Chest 116, 968.
[2] HB Burke, PH Goodman, DB Rosen, DE Henson, Weinstein JN, Harrell FE Jr, Marks JR, Winchester DP, Bostwick DG, 1997, Cancer 79, 857.

Figure 8.4. The MyFirstNEURON start window

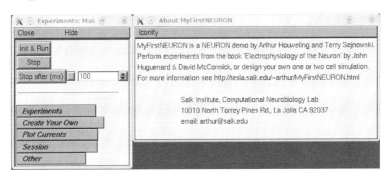

1. Run MyFirstNEURON.[3] MyFirstNEURON should start with two screens similar to those shown in Figure 8.4, "The MyFirstNEURON start window".

2. In the "Experiments: Main control" window, press the "Experiments" button and select "Basics" and select **Resting potential**. This loads parameters which simulate a cell membrane containing both potassium and sodium leak channels; it should also bring up two additional windows: "Graphs" and "Parameters".

3. The *initial* ion concentrations are described under the "Concentrations (mM)" section of the "Parameters" window. For example, ki (mM) and ko (mM) represent the concentrations of potassium inside and outside of the cell, respectively. **Are the intracellular and extracellular concentrations for potassium and sodium within the ranges typical for a mammalian neuron?**

4. Press "Init and Run" (in the "Experiments: Main control" window). What is the membrane potential generated across the cell membrane? **Does it change with time? If you answered in the**

affirmative, what is the range of change in potential you observed during the course of the simulation?

5. Change the external potassium concentration to 135 mM. Evaluate this change (press "Init and Run"). **What is the membrane potential generated across the cell membrane? Is this expected? Why?** Change the external potassium concentration to 0.1 mM. Evaluate this change. **What is the membrane potential generated across the cell membrane?** Reset all values to the default values (click on the checkmark next to the parameter(s)).

The lower portions of the right-hand column of the "Parameters" window (Figure 8.5, "The MyFirstNeuron 'Parameters' window") displays parameters associated with externally imposed currents and voltages across the membrane. The membrane can be held at a fixed voltage and current across the membrane measured or, alternatively, a fixed current can be forced to flow across the membrane and the voltage measured. With "current clamping", the amount of current passing across the cell membrane can be fixed at a constant value. In the next set of experiments, you will explore the factors (e.g., current amplitude, ion concentrations) which influence the triggering of an action potential. The model neuron we are evaluating consists of a single compartment with a membrane surface area of 29000 μm^2 and a membrane capacitance (cm) of 1 $\mu F/cm^2$. This model neuron also contains "fast" voltage-gated Na^+ and K^+ channels which are responsible (in part...) for action potential generation.

Figure 8.5. The MyFirstNeuron 'Parameters' window

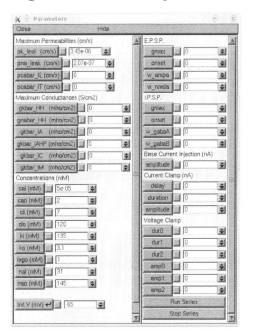

You may find it useful to remember the fundamental relationship you learned in physics regarding voltage and current, Ohm's law (V/I = R where V = voltage, I = current, and R = resistance). Often the word "potential" is used to refer to a voltage.

1. In the "Experiments: Main control" window, press the "Experiments" button and select "Basics" and select **Membrane properties**. This loads parameters which simulate a cell membrane containing potassium and sodium leak channels, but also with an externally imposed current being held across the membrane (the current clamp "delay" parameter is now set to 10 and the "duration" parameter is now set to 50). Try pressing "Init & Run". *What happens to the membrane potential at time 10 (when the current is imposed)?*

2. Follow the membrane potential and move the amplitude of the current clamp up step-wise (0.1 or 0.05 step magnitude) from 1.6 nA. **_Estimate the threshold potential of the Na^+ channels in the model neuron to within ± 2 mV._** Print a window showing the action potential firing when you have increased the current clamp sufficiently.

Exercise: constructing a synapse

The NEURON software package can also be used to simulate signaling in networks of neurons. The items in monospaced fonts below represent buttons to push or pull-down menu selections to select.

First, start NEURON (in linux, execute nrngui from the command line). This should bring up the main menu window of NEURON (Figure 8.6, "The NEURON main menu window").

Figure 8.6. The NEURON main menu window

Define the cell types

1. Define a cell that can supply afferent signals (physiologically, an afferent cell is a cell carrying information *toward* the central nervous system). Note that items in this font represent buttons to press on the NEURON interface. For example, Build-Network Cell represents an instruction to click the 'Build' button and then select or click the 'Network Cell' button.

`Build-Network Cell-Artificial Cell`

- this should bring up the the ArtCellGUI window (Figure 8.7, "The ArtCellGUI window")

`New-NetStim` to set the parameters for this cell:

- set **number**: specify a 'train' of 10 events
- set **interval**: events occur at intervals of 10 ms
- set **start**: the train starts at t=50 ms

Rename

give cell a name (e.g., "AffCell")

Figure 8.7. The ArtCellGUI window

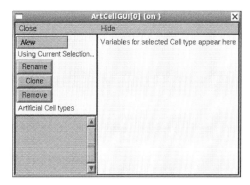

1. Define a cell that can integrate synaptic inputs

   `New-IntFire 1`

   - an IntegrateFire object has an "excitation" (analogous to membrane potential) that decays toward 0 with time constant **tau**; set tau to 10 ms

- **refrac** defines the "refractory interval", the time interval after firing in which cell is unresponsive to inputs; set refrac to 5 ms

  Rename

  - give the cell a name (e.g., "IntCell")

2. Save the ArtCellGUI tool to a session file (don't close the window!). This allows you to load the session and start from this point if you want to 'start over' at some point later.

   In NEURON Main Menu window,

   File - save session

   note: don't try and put the file name in the filter space

Define a network

Place the cells in the network:

- Build - NetworkBuilder

  - note the instructions on the right of the panel:

    - a cell can be created by dragging from the list on the left onto the whitespace

    - a cell can be replaced by placing a new cell over it; a cell can be deleted/discarded by dragging the cell off of the view

- Create a cell to generate an afferent spike train:

- drag an "AffCell" onto the space

- Create an "integrate and fire" cell:

  - drag a "IntCell" onto the space

- Create the synapses:

  - Click the `Src->Tar` radio button

  - Drag from AffCell to IntCell (look for changes in the font color of the cell label to ensure you have actually selected a source and target cell)

- Consider saving the session to a session file so you can start at this point by loading the session file if you want to 'start over' at some point later.

Network simulation

1. Create the network: click on the `Create` checkmark box in the "NetGUI" window

   - note that this is a 'point of no return' (you may not be able to modify the network and then click 'create' again - if you want to modify the network connections, you may need to exit and restart NEURON)

2. ***Document the network: print a copy of the "NetGUI" window*** (see Appendix F, *Screenshots*)

3. Bring up additional windows to aid in visualizing network activity:

- Click the SpikePlot button to bring up a window where the input and output spike trains can be visualized

4. Run the simulation: in the NEURON Main Menu window,

   Tools - RunControl

   - **Tstop** defines the the stop time for the "Init & Run" routine; set Tstop to 200ms

   Init and Run

Analyzing data from the simulation

Figure 8.8. The NEURON spike plot window

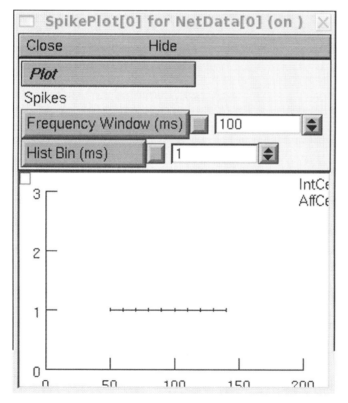

Neural networks

1. At this point, you may recognize that the legend for the spike plot window (Figure 8.8, "The NEURON spike plot window") could be more informative. While, there is a "spike train", which cell does the train represent? To make sense out of things,

   - right click in the SpikePlot window

   - select Color Brush

   - select a distinct color

   - left click on the spike train of interest

   - note that the legend now reflects the new color

2. ***Record your observations from the "SpikePlot" window (e.g., print a copy of the window). Ensure you number the figure and refer to that number when answering this question. Ensure you clearly label each spike train in the figure, so that it is apparent which spike train is associated with which cell. How do you interpret the data?***

3. Network building: define the strength of synapse connections

   Close the spike plot

   Click the Weights button in the "NetGUI" window

   - a weight of 0 means the presynaptic cell does not influence the postsynaptic cell at all

Neural networks

Click the connection of interest (there should only be one synapse at this point...) on the list on the right

- this should bring up the NetEdgeGUI window

- the weight shown in the white window is the weight of the synaptic connection selected in the right-hand column

Set the weight to 2. It works like this:

- if you click on a synapse ('edge') in the right-hand column/panel, its weight will be set to the weight indicated in the "Weight" window at the top of the screen

- if you select a number from the left panel, its value will be placed in the field editor

It's okay to close the weights window when you're finished with it.

4. Try clicking the "Create" button. You may be told you can't proceed. If this is the case, save your work (save the session), exit NEURON, start NEURON, load your session.

5. After checking the "Create" button, press the Spike Plot button. Press the `Init & Run` button in the RunControl window.

6. **Record your observations. Do the spikes associated with the cell downstream of the afferent cell occurr at the same time as the**

spikes from the afferent cell? Why are the spikes be arranged this way?

Exercise: simple pattern recognition

Obviously, a two-neuron artificial network like the one constructed above, doesn't exactly have the power of your CNS (we hope!). In the next exercise, we will attempt to mimic how a CNS might recognize a simple pattern. A perceptron is an artificial neural network consisting of a set of input neurons, an output neuron, and, optionally, one or more intermediate "layers" of neurons.

Digital circuits such as the chips which are used in a cell phone or a computer are composed of large numbers of digital logic gates, electronic representations of Boolean logic functions such as AND, OR, NOT, and XOR[4].

Using the skills you learned in the previous exercise, create a perceptron which functions as the equivalent of an "AND gate". This perceptron analyzes binary two-digit numbers and **specifically** recognizes the signal "11" (i.e., distinguishes 11 from 00, 01, and 10). To construct this network, "training" will be required: known patterns must be presented to the network and the weighting adjusted appropriately until the weighting pattern can be left fixed with correct pattern recognition occurring. Some steps to help you get started:

---

[4]XOR represents the "exclusive or" operator: in a nutshell, False XOR False is False; False XOR True is True; True XOR False is True; True XOR True is False

Neural networks

1. Mentally prepare yourself. If a neuron is analyzing a signal, what do you define "recognition" as? For example, if a signal is presented and a spike train is generated by the processing neuron, does this constitute "recognition"? What if a signal is presented but the processing neuron doesn't fire at all? Does this constitute "recognition"?

2. In preparation for defining the input signal, define a NetStim cell and name it ONOFF1. Define a second NetStim cell and name it ONOFF2. These cell types will serve as the source of the "0" (off) or "1" (on) signals.

3. Define a cell called INTFIRE1 that can integrate synaptic inputs (as described in the first exercise); this is the cell that you hope to have process the signal in a specific fashion.

4. Use the "Network Builder" tool to create a network which simulates a "11" signal (two OFFON cells) being processed by a single neuron.

   - place a ONOFF1 and a ONOFF2 cell on the building space

   Note:

   - The OFF signal should have an interval of 1000 ms (a "0" signal corresponds to an afferent neuron which sends a signal train with very large spacing between spikes). A "1" signal corresponds to a rapidly firing neuron.

- The ON signal should have an interval of 10 ms. (a "0" signal corresponds to an afferent neuron which sends a signal train with very large spacing between spikes)

- place an INT cell (the cell processing the signal) on the network

- create a synapse between each ONOFF cell and the INT cell

- weight each synapse with a value of 2

- use the ArtCellGUI to set the interval of each OFFON cell to a value of 10 (corresponding to each OFFON cell sending a "on" signal)

- run a simulation of the network

    - check the `Create` checkbox

    - click on the `SpikePlot` button in the "NetGUI" window

    - click on the `Init & Run` button in the "RunControl" window after setting Tstop to 200

What is the output of the system (of the INT neuron)? (print the output) How do you interpret this data (i.e., is the INT cell "recognizing" the "11" signal?)?

5. Check to see if the system recognizes other patterns correctly:

- e.g., to represent a "10" signal, use the ArtCellGUI window to change the firing interval of ONOFF1 to 10 and ONOFF2 to 1000

- run a simulation of the network:
  - click on the SpikePlot button in the "NetGUI" window to generate a new spike plot for the new set of cells
  - click on the Init & Run button in the "RunControl" window after setting Tstop to 200

  What is the output of the system (of the INT neuron)? (print the output) How do you interpret this data?

6. Try the other possible patterns (note that you must define synapses and weights *every* time you modify the network as well as generate a new SpikePlot window every time you add or remove cells from the network). **In each case, what is the output of the system (print and interpret the output)?**

7. **Does this system need more "training" or does the initial weighting scheme provide acceptable recognition? How are you defining "acceptable recognition"?**

8. **Are there modifications you can make to the weighting to cause the system to specifically recognize its target (i.e., which weight values provide maximal specific recognition of 11?) What are you defining as "specific**

recognition"? Does the system give an unambivalent strong signal for 11 and an unambivalent weak signal for 10, 01, and 00?

9. ***How do you think it would be possible to enhance discrimination between 11 and the other two-digit binary numbers? Propose a different network design providing more specific recognition.***

10. Optional question to consider: Would it be challenging to develop a network which recognizes numbers (expressed in binary) which are multiples of two? Contrast the difficulty of developing the first network with the difficulty of developing a network which recognizes numbers (expressed in binary) which are multiples of three. Explain.

11. ***How might 'training'/'learning' occur in a biological neuron network?***

12. ***How many neurons and synapses do you think would be required to recognize a more complex structure like a human face (e.g., distinguishing between a human and an animal face)? Would more neurons and synapses be required to specifically recognize individual human faces (i.e., to differentiate between Joe, Bob and Bill)?***

Installing and running NEURON

NEURON is a software package, written by Michael L. Hines (Yale University), capable of simulating some of the electrophysiological aspects of a neuron. To use

NEURON, you need to have access to a computer which has **NEURON** installed on it. You can reach this goal by using a computer with **NEURON** already installed, by installing **NEURON** on the computer you plan to use for this exercise, or by using a "live CD" with **NEURON** on it.

To install NEURON (version 5.1 or newer) on a computer:

- download the NEURON software onto your computer (Linux, Microsoft, and MacOS X versions are available at at the NEURON web site [http://www.neuron.yale.edu])

- follow the installation instructions for your platform

Running NEURON:

If you are using Linux, type `nrngui` at a command prompt.

Installing and running MyFirstNEURON

The MyFirstNEURON package was developed by Arthur Houweling and is available at [http://www.cnl.salk.edu/~arthur/MyFirstNEURON.html]. A more recent version may be available at [http://senselab.med.yale.edu/modeldb] (use accession number 3808).

MyFirstNEURON requires NEURON to be installed.

Linux:

1. Unzip the MyFirstNEURON zip file. Create a directory named myfirstneuron, move the zip file into the directory, then unzip the file.

2. Run nrnivmodl (an executable in the neuron package) *in the directory which contains the MyFirstNEURON "mod" files (wherever you unzipped the MyFirstNEURON zip file)*. Running nrnivmodl will create an executable called "special" in a directory labeled i386, i686 or something along those lines.

3. At this point, from the directory which contains my1stnrn.hoc (wherever you unzipped the MyFirstNEURON zip file), execute ./i686/special my1stnrn.hoc - (replacing i686 with the appropriate path, if necessary). The "-" at the end is important; otherwise the program will start and finish immediately. This should run NEURON and bring up a set of several windows.

4. Use CTRL-D to exit the neuron program when finished.

Mac:

Best wishes. While NEURON will run on a Macintosh, we were unable to run MyFirstNEURON on a Macintosh (as of 2007).

Windows:

- If you are using Windows, note that if you installed MyFirstNEURON using the my1stnrn.nrnzip link on the web page, it should automatically execute the program. If it does not, try:

    - drop the my1stnrn.nrnzip file on the mos2nrn application

- double-click on "my1stnrn.hoc" (located in the folder "nrn0t.***" installed during MyFirstNEURON installation)

- Some windows computers may complain about a / tmp directory not being available. If this concerns you, make the directory (via the MS-DOS prompt or using the file explorer).

Readings

Holland, J, 1999, Emergence: From Chaos to Order, New York: Basic Books.

- a relatively accessible discussion of how complex behaviors can 'emerge' from a system governed by a relatively simple set of rules; includes a treatment of neural networks

Koch C, 1999, Biophysics of Computation, New York: Oxford University Press.

Ullian EM, KS Christopherson, and BA Barres, 2004, Role for Glia in Synaptogenesis, Glia 47, 209.

Huttenlocher PR, 1979, Synaptic density in human frontal cortex - developmental changes and effects of aging, Brain Res 163, 195.

Huttenlocher, PR and AS Dabholkar, 1997, Regional differences in synaptogenesis in human cerebral cortex, J Comp Neurology, 387, 167.

Nimchinsky EA, BL Sabatini, and K Svoboda, Structure and function of dendritic spines, Ann Rev Physiol 2002. 64:313–53

Chapter 9. Protein concentration

Protein concentration analysis

- in advance, read the protocol and prepare a detailed table describing how you will prepare each sample for the Bradford assay

- see Appendix G, *Protein assays* for an overview of protein concentration assays

Procedure

prepare the sample

If the samples are muscle tissue samples, they must be prepared prior to analysis:

- cut the tissue sample into 8 cubes of appr. 0.25 to 0.5 cm per side

- place the cubes in a 1.5 mL snap-cap centrifuge tube

- to the 1.5 mL tube, add 240 µL 1X SB supplemented with β-mercaptoethanol[1]

- flick the snap-cap centrifuge tube 15 times with your finger to mechanically disrupt the tissue and to suspend the tissue in the buffer (alternatively, briefly vortex the sample)

- incubate the sample for 5 min at room temperature

- decant the buffer (but *not* the solid pieces of tissue) into a 1.5 mL screw-cap tube

note: you should not be concerned about getting every last drop of buffer since you will probably use no more than 30 µL of the sample

- heat the decanted liquid at 95°C for 5 min

- if you are not ready to proceed to the next step, you may store the extract no more than 3 h at room temperature (alternatively, you may store the extract for several weeks at -20°C)

determine the concentration of protein in the sample

1. Determine the protein concentration of your sample(s) using bovine serum albumin as a standard in conjunction with the Bradford assay (see the detailed protocol, below). Prepare all samples and standards for the Bradford assay in duplicate.

2. Ensure that, after completing the protein concentration assay, you save the original sample (label it and store it in the -20°C freezer) if it will be used in a future laboratory session.

Laboratory report

Unless the instructor indicates otherwise, a single report should be turned in describing the data both from this week and from preceding and succeeding labs which also evaluate the sample(s). Evaluating this week's data is important in preparing for next week's lab. For this week's data, consider the following:

1. Calculate the estimated concentration in your protein sample: first, generate a standard curve using a linear or non-linear regression as you deem appropriate; second, use the curve to determine the protein concentration in each dilution of the sample; third, use the dilution factor to calculate the estimated concentration of the original sample.

2. Does the standard curve look "good" or are there unexpected aberrations in the location of the data points?

3. Was the variability between your replicate samples significant? what about the variability between the estimated concentration using different sample dilutions? what might contribute to the variability in each case?

Introduction

The Bradford total protein quantitation assay [bradford1976], and several commercial modifications thereof, are colorimetric assays based on the tendency of the dye Coomassie G-250 to shift absorbance from 465 nm to 595 nm (there is a simultaneous color change of the reagent from red/brown to blue) when the reagent binds proteins in an acidic solution. Although it is common to simply use absorption at 450 as the endpoint of the assay, the sensitivity and linearity of the assay can be improved by evaluating the ratio of absorption at 595 to absorption at 465 for each sample (Zor and Selinger 1996). The mechanism of the reaction is based on an anionic form of the dye; this form interacts primarily with arginine residues and, more weakly, with histidine, lysine, tyrosine, tryptophan, and phenylalanine residues. The reaction reaches a relatively stable endpoint so valid absorption measurements can be made over the course of the hour following equilibration of the dye and protein.

The Bradford assay does not have a wide linear range. Thus, it may be necessary to prepare sample dilutions prior to analysis.

The Bradford assay is subject to interference from some compounds. The following compounds are not likely to interfere at the indicated concentrations:[2,3,4]

β-mercaptoethanol, 1M

dithiothreitol, 5 mM - 1M

EDTA, 0.1M

EGTA, 0.002 - 0.05M

ethanol, 10%

glucose, 20%, 1M

HEPES, 0.1M

SDS, 0.1%

Triton X-100, 0.1 - 0.125%

urea, 3 - 6M

---

[2] Bio-Rad, Bio-Rad Protein Assay Rev C
[3] Thermo Scientific, 2009, Thermo Scientific Pierce Protein Assay Technical Handbook
[4] unpublished data

Protein concentration

In any protein assay the ideal standard is a purified preparation of the protein or protein mix being assayed. Since color yield varies with protein with the Bradford assay, the choice of this standard is significant if precise quantitation is desired. However, such a reference standard often isn't available. In such a situation, another protein (or set of proteins) is used as a relative standard. The ideal relative standard is that protein or protein mix which gives a color yield similar to that of the protein being assayed. Selecting such a protein standard is generally done empirically. Alternatively, if only relative protein values are desired, any purified protein may be selected as a standard.

Procedure

1. If BSA stock solutions are not provided, prepare BSA (bovine serum albumin) stocks as described below:

    3 mg/mL BSA

    example: add 30 mg BSA to 10 mL water

    at least 20 µL of 1 mg/mL BSA

    example: add 16.7 µL of 3 mg/mL BSA solution to 33.3 µL water

    at least 30 µL of 0.1 mg/mL BSA

    example: add 3 µL of 1 mg/mL BSA solution to 27 µL water

2. Prepare (label) a set of tubes according to Table 9.1, "Samples and standards"

Protein concentration

3. Add water to *each* tube following Table 9.1, "Samples and standards"[5]

4. Add protein (sample or standard) to *each* tube according to Table 9.1, "Samples and standards"

5. Add 200 µL dye ('Bradford reagent') to each tube; vortex vigorously; decant into plastic 1 mL cuvets

6. Wait at least 10 min (but not more than 1 h) at room temperature after adding dye

7. Zero the spectrophotometer (see Appendix H, *Spectrophotometer use*) with water at A_{590}

    - after zeroing at A_{450}, record the baseline A_{450} value (the value with water) so that subsequent A_{450} measurements can be blank-corrected

8. Measure A_{590} and A_{450} for each sample:

    - directly before measuring, vortex or invert two or three times to mix the sample

    - decant the sample from the microcentrifuge tube into a cuvet

    - record each absorbance value

Protein concentration

Table 9.1. Samples and standards

Standard (μg/mL)	μL standard	μL water	μL dye	A_{590}/A_{450} (1)	A_{590}/A_{450} (2)
0		800	200		
0.2	2 (0.1 mg/mL)	799.8	200		
1	10 (0.1 mg/mL)	799	200		
5	5 (1 mg/mL)	795	200		
10	10 (1 mg/mL)	790	200		
60	20 (3 mg/mL)	780	200		
Sample					
0.1	0.1	799.9	200		
2	2	798	200		
8	8	792	200		

calculations

1. Calculate A_{590}/A_{450} for each sample

2. Construct a standard curve

3. Use linear regression to mathematically define the observed relationship between protein concentration and A_{590}/A_{450}

4. Evaluate the concentration of the unknown sample(s)

Chapter 10. SDS-PAGE

- **in advance of this lab session, you must prepare a "loading table" (see the "Planning" section below)**

- this experiment will require the full laboratory session time and, perhaps, additional time outside of the scheduled lab

- polyacrylamide is a neurotoxin; it is a good idea to wear gloves when handling polyacrylamide gels

Introduction

Electrophoresis is a powerful means of separating molecules. Sodium dodecyl sulfate-polyacrylamide gel electrophoresis (SDS-PAGE) is frequently used to separate polypeptides on the basis of their molecular masses. In this technique, prior to electrophoresis, proteins are denatured by heating in the presence of the strong detergent, SDS. A reducing agent such as β-mercaptoethanol or dithiothreitol may be added to disrupt disulfide bonds. In this state, the number of SDS molecules bound to the polypeptide is approximately proportional to the mass of the protein. When subjected to an electric field in a polyacrylamide gel, these polypeptide-SDS complexes migrate with a velocity roughly proportional to their mass.

SDS-PAGE is typically used for protein analysis; variants of PAGE are also used for preparative work. SDS-PAGE can be used as a second dimension, in conjunction with isoelectric focusing, to effect a two-dimensional protein separation. However, probably the most frequent use of SDS-PAGE is in conjunction with Western blotting: after subjecting a protein mixture to electrophoresis, the polypeptides in the gel are transferred to a membrane, either electrophoretically or via capillary action, and the membrane incubated with antibodies against a specific protein. Detection of the antibodies bound to the protein on the membrane serves as a highly specific and sensitive means of detecting a particular protein of interest, and of quantitating levels of that protein.

In this lab you will subject a sample containing (hopefully!) protein to SDS-PAGE. You will evaluate the results either by staining the SDS-PAGE gel or by Western blotting.

Procedure

planning

Before the lab session, plan how you will "load" the gel. "Loading" the gel refers to the process of adding a predetermined volume of each sample to a well in the gel. Each sample is added to a separate well (i.e., one well per sample). To load the gel, one must plan ahead and first determine the sample mass, and from that value, the corresponding sample volume, that will be loaded in each lane. In this exercise, 10 µg of each sample should be loaded. Prior to loading, each sample is heated in a solution of SDS and β-mercaptoethanol in order to completely denature the sample.

SDS-PAGE

Use the information provided below along with the protein concentration estimates you obtained from the Bradford analysis of your unknown sample(s).

In addition, you will be provided with a pre-stained mass standard "ladder". For this sample, your instructor will indicate the volume of the sample which should be loaded. An example of a pre-stained molecular mass standard is Lonza's "ProSieve Color Protein Markers" mixture. This mass ladder containing polypeptides with apparent masses of 181, 121, 77, 48, 39, 25, 19, 12, and 10 kDa. The bands are prestained with different colors in order to aid the user in identifying the bands (the colors corresponding to the bands described above are purple, purple, red, red, purple, red, red, purple, and red, respectively.

SDS-PAGE

Figure 10.1. A SDS-PAGE gel

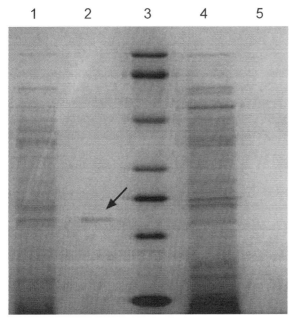

Lanes 1 and 4 represent electrophoresis of crude cellular extracts and lane 3 represents electrophoresis of a molecular mass protein standard mix (top to bottom: 90, 66, 45, 34, 27, 20, 14.4 kDa)

calculating loading values

To prepare to load the gel, you must make several calculations. For example, consider a gel in which you plan to load 10 µg of 'sample 23' in lane 8:

1. To add 10 µg of the sample, you must know the protein concentration of the sample (either determined by a protein assay you conducted or provided by the instructor). *If you already did a Bradford assay of an unknown, <u>you</u> should use <u>your</u> data for this calculation.* 'Sample 23' has an estimated concentration of 2.5 µg/µL. Given this, how many µL of sample correspond to 10 µg? This

SDS-PAGE

volume is the volume of sample you will place in the tube in which you prepare the final solution to be loaded onto the gel. This is also the value which you should enter in your loading table in the column labeled 'volume sample/µL'.

2. The sample should be adjusted to 1X sample buffer (1X SB) and the final volume of the sample to be loaded should be 15 µL. Therefore, the amount of $3X^1$ sample buffer which should be added is 15 µL / 3. What is 15 µL / 3 ? This volume is the volume you should enter in your loading table in the column labeled 'volume 3X SB / µL'.

3. Calculate the additional volume of water required to bring the final volume to be loaded to 15 µL. If you have added 5 µL of 3X SB and 2 µL sample, how many more µL will you need to add to end up with a final volume of 15 µL? This amount should be entered in the column of the loading table labeled 'volume H_2O / µL'.

These calculations should be made for each gel lane to be loaded. (Just in case you didn't catch it, you should organize the calculated values into a table). For example,

Table 10.1. Example of a single row from a gel loading table

lane	sample	µg sample	µg/µL sample	µL sample	µL water	µL 3X SB	µL total
8	unknown	10	2.5	4	6	5	15

thinking ahead: some technical notes on SDS-PAGE

If you use SDS-PAGE in the future, you may find these pointers helpful in obtaining better results

- the best resolution is usually obtained with approximately 1 µg protein or less

- the sample loading volume should be less than 1/2 the height of the sample well

- for best results, load 1X SB into any sample wells that are not loaded with sample

preparing the samples

Figure 10.2. A screwcap tube and cap

Prepare samples (unless directed otherwise, use a 15 µL loading volume)

1. Determine the sample volume, sample buffer volume, and amount of water that should be added to prepare the final sample to be loaded. If you skipped the guidelines given above, now is the time to go back and read them more closely.

2. Obtain a 1.5 mL screw-cap tube (Figure 10.2, "A screwcap tube and cap") for each sample (screw-cap tubes are used to ensure that lids don't "pop open" during heating).

SDS-PAGE

3. Directly before use, add β-mercaptoethanol (β-ME) to the 3X sample buffer solution. Prepare 3X sample buffer supplemented with β-ME by adding 5 µL of β-ME to 95 µL of 3X sample buffer. Use this modified 3X buffer in preparing your samples.

4. Prepare your samples in 1X SB (by adding the appropriate volumes of 3X buffer and water) using the loading table to guide you. Mix thoroughly. Keep samples on ice.

5. Denature your samples (prepared in 1X SB in the previous step) by heating for 5 min at 95°C in a heat block or in a water bath. Cool for 2 min on ice.

- Check with the instructor to determine whether you *should* or *should NOT* heat or denature the prestained mass marker sample (if such a sample is being used)

- It's probably not such a bright idea to immerse the sample tube in liquid during cooling (in at least one instance, students have discovered physics hard at work as liquid is 'sucked in' to the the tube as its contents cool).

preparing the gel electrophoresis apparatus

Figure 10.3. PAGE gel electrophoresis box

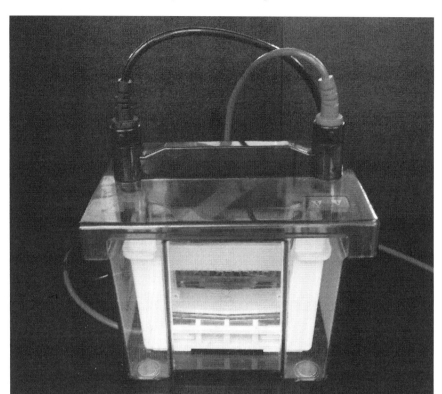

Rinse the gel electrophoresis box (Figure 10.3, "PAGE gel electrophoresis box")

Figure 10.4. SDS-PAGE gel

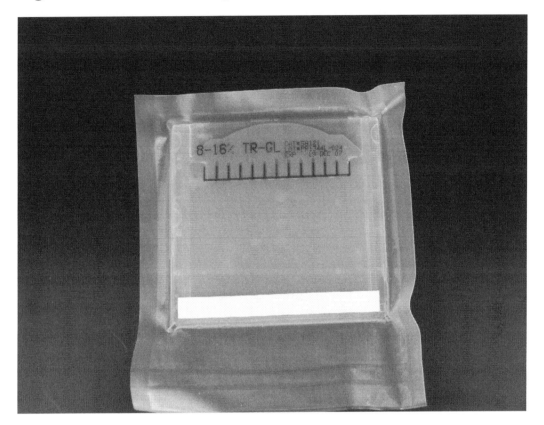

Obtain the SDS-PAGE gel (Figure 10.4, "SDS-PAGE gel"):

For Lonza PAGEr gels and Cambrex Duramide gels, cut open the pouch and remove the gel. Peel the tape off the bottom of the cassette.

For Jule "Snap-A-Gel" gels, cut the envelope containing the gel and remove the gel; snap off the bottom tab on the gel.

Carefully remove the comb from the gel by sliding it with a slow and steady motion straight up and out from the cassette.

SDS-PAGE

Rinse wells with 1X running buffer to remove unpolymerized or partially polymerized acrylamide fragments, etc. -- this can be done (1) with a squirt bottle filled with running buffer or (2) with a syringe and needle. Be gentle, you don't want to damage the gel.

Using a permanent marker, draw lines underneath each well on the outside of the tall plate so that it will be easier to see well locations when loading the samples.

Assemble the gel electrophoresis apparatus:

Place the gel cassette into the electrode assembly with the short plate facing inward.

note:[for the MiniProtean III, make sure the gel is forced all the way down so the two blocks at the bottom hold the gel in place (this may take a bit of force -- ask the instructor if you would like assistance)]

Slide the entire assembly and cassette into the clamping frame.

note:[make sure the rubber gasket (Figure 10.5, "Gel gaskets") fits snugly against the gel - you may have to switch the gasket to the other side (the gasket should be placed so the flat side faces outward for Duramide gels) to get a tight seal]

Figure 10.5. Gel gaskets

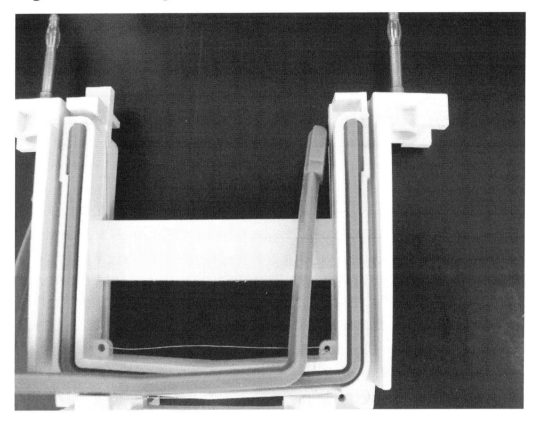

Press down the electrode assembly while closing the two cam levers of the clamping frame.

Lower the inner chamber into the "mini tank".

Fill the inner/upper chamber of the apparatus with 1X running buffer. Before proceeding, ensure that no leaking occurs.

Load samples:

Using a gel loading tip, withdraw sample from tube using a 20 µL pipettor *or* use a 10 or 20 µL syringe.

Ensure there are no bubbles or air in the sample.

SDS-PAGE

Insert the tip into a well so that it is approximately 4 mm from the bottom of the well.

Slowly depress the plunger of the pipettor (so that the sample collects at the bottom of the well).

Carefully layer running buffer on top of the wells.

Ensure that the bottom electrophoresis chamber is filled with 1X running buffer (the upper and lower chambers together will require approximately 350 mL 1X running buffer). If you have sufficient buffer (and if you are certain your apparatus is not leaking), fill the chamber to the bottom of the sample wells to keep the gel cooler.

Run the stacking gel at 100 V (150V if using a Gradipore gel) until protein bands have passed from stacking gel into separating gel at least 1 cm. This should take 45 min to 1 h.

Run the separating gel at 180 V (run until the dye front is approximately 1 cm from the bottom of the gel). This should take approximately 1 h.

Turn off the power supply and remove the top of the gel electrophoresis apparatus.

Remove the gel holder; remove the gel (and the two plates sandwiching it) from the holder.

Using a plastic or metal lever, twist the edge of the lever between the edges of the two plates so that the two plates separate.

Remove one plate slowly, allowing the gel to stick to the other plate.

At this point, ensure that you understand whether you will be staining the gel or performing a Western blot with the gel.

staining

Fill a glass tray at least 1 cm deep with the gel staining solution.

Allow the gel to be drawn into the tray by capillary action (it should "peel off" the plastic plate).

Staining (colloidal Coomassie technique):

Fix the gel: immerse the gel for (1) 30 min w/gentle rocking in 1X fixing solution (12% trichloroacetic acid, 3.5% (w/v) 5-sulfosalicylic acid) or (2) 60 min in 50% methanol/10% acetic acid.

Obtain fresh stain solution.

Immerse the gel in the staining solution; seal container tightly; rock gently for a minimum of 1-2 h (best sensitivity is with overnight staining - background rises after appr. 12 h staining).

Briefly wash in 20% methanol (aq) (for really clear background, destain in water several hours or overnight).

Record the results (a low-end scanner works fine for this).

For stable storage (over 3 d) (1) store in 25% w/v ammonium sulfate rather than water and store at $4^{\circ}C$ or (2) rinse very thoroughly with water and place between cellophane sheets to dry.

SDS-PAGE

Laboratory report

The laboratory report should follow the format described in the first laboratory session and in the course syllabus. This laboratory report should include data from both the previous laboratory session and this laboratory session.

Evaluate your data:

Characterize the nature of separation produced by SDS-PAGE: do your results support the assertion that the distance migrated is proportional to molecular mass? Is there a linear correlation between distance migrated and mass?

Evaluate the molecular mass(es) of the protein component(s) of the unknown sample by generating a calibration curve using the molecular mass standard(s) or the protein ladder. Note that here we are referring to a set of standards which correspond to proteins of known mass (e.g., 10 kDa, 30 kDa, 50 kDa, ...).

If you used mass loading standards (versus a conventional protein ladder), evaluate the concentration of the unknown protein by generating a mass calibration curve using the mass loading standards. Note that here we are referring to a set of standards which correspond to known masses of protein loaded (e.g., 0.1 µg, 1 µg, 5 µg, ...).

See also Appendix I, *Densitometry*.

Some resources

SDS-PAGE

molecular mass determination

After staining molecular mass standards by a suitable method (if they are not prestained), a reference graph can be made by plotting the relative mobility, R_f,[2] of each standard versus the corresponding molecular mass. The relationship is approximately exponential: plotting R_f versus log(molecular mass) should yield a linear plot. The molecular mass of unknowns can then be estimated from the relationship described in the above plot.

sample concentration

Densitometry, the quantitation of optical density in a light-transmitting material, can be used to estimate the mass of protein present (i.e., how much protein was actually loaded in a given lane). This allows SDS-PAGE to function as a quantitative tool for estimating protein concentrations, providing data which parallels the estimates obtained using the Bradford assay. Some software packages with densitometry capabilities are listed in Appendix I, *Densitometry*.

Pre-lab notes

The sample that will be used in this lab will be described by the instructor prior to the lab.

Unless the instructor gives you protein concentration data, you <u>must</u> analyze the protein concentration data you obtained in earlier labs to calculate the protein

---

[2] the distance traveled by the protein of interest divided by a reference distance (e.g., the distance traveled by a tracking dye)

concentration of your sample. This concentration estimate is required to prepare for this lab.

Chapter 11. Western blot analysis

Western blot

Probably the most frequent use of SDS-PAGE is in conjunction with Western blotting: after subjecting a protein mixture to electrophoresis, the polypeptides in the gel are transferred to a membrane, either electrophoretically or via capillary action, and the membrane incubated with antibodies against a specific protein. Detection of the antibodies bound to the protein on the membrane serves as a specific and sensitive means of detecting a particular protein of interest, and of quantitating levels of that protein.

In this lab, a sample containing (hopefully!) protein is subjected to SDS-PAGE and then a 'Western blot' is performed, transferring the proteins in the gel to a membrane. Proteins on such a membrane can be detected either in a non-specific fashion, using a stain such as India ink, or in a specific fashion, using a primary antibody against the protein of interest along with a system for visualizing the antibody on the membrane.

Procedure

advance preparation

While SDS-PAGE is underway, ensure that the transfer apparatus and transfer buffer are prepared and that you understand how to assemble the apparatus. Items to have on hand:

- transfer module (big plastic 'box' which holds everything else)

- transfer cassette (little plastic 'holder' which fits in module and holds fiber pads, membrane, and blotting paper as a 'sandwich')

- fiber pads (2)

- membrane (1)

- blotting paper (2)

- cooling accessory

- power supply

- test tube

- a tray which can be used to submerge the cassette, paper, pads, and membrane in transfer buffer

Figure 11.1. Open transfer cassette with fiber pads and blotting paper

transfer

Rinse the transfer apparatus.

Soak the transfer pads thoroughly in transfer buffer.

Mark the membrane in a corner with your initials. This mark facilitates:

- identifying the membrane later

- identifying the orientation of the membrane relative to the gel (i.e., (1) which side of the membrane faced the gel during the transfer and (2) which side of the membrane corresponds to the first lane in the gel)

Incubate the membrane in transfer buffer.

Incubate the blotting paper in transfer buffer.

After SDS-PAGE is complete, incubate the gel in transfer buffer for 15 min.

Assemble the 'transfer sandwich':

- add transfer buffer to a container (e.g., a glass tray) to a depth sufficient to cover the membrane when the 'sandwich' is assembled and place the transfer cassette in the container with the black side down

    - place an open cassette in the tray, black side down

    - lay a fiber pad on the black side of the cassette

    - lay one wet blotting paper on the fiber pad - use a test tube to roll out any air bubbles

- lay gel squarely on the blotting paper - use a test tube to roll out any air bubbles

- lay the membrane on the gel, *noting the orientation of the gel relative to the mark you made earlier on the membrane* - use a test tube to roll out any air bubbles

- lay one wet blotting paper on the gel - use a test tube to roll out any air bubbles

- lay a fiber pad on top of the paper

Figure 11.2. Schematic of assembly of Western transfer

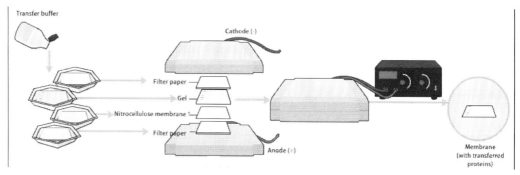

Close the cassette and securely clamp it together; immediately place the cassette in the transfer unit so that the black side faces the negative electrode of the unit (the membrane should be closer to the positive electrode and the gel should be further from the positive electrode unless using acidic conditions).

Add a frozen ice container to the unit or connect water lines to cooling module.[1]

---

[1] optional at 20V but strongly recommended at higher voltages

Fill the unit with transfer buffer so that the transfer buffer is at least level with the top of the transfer cassette.

Place a stir bar in the tank.[2]

Place the lid on the tank and ensure the power cords are (correctly) attached to a power supply. Blot at 20 V for 2.5 h.

notes:

- you may leave after blotting begins; the blots can remain submerged in the tanks at room temperature overnight after the transfer is finished

  - once the transfer apparatus is disassembled, blots can be stored:

    - in blocker at $4^{\circ}C$ for up to one week (Bio-rad)

      - for extended periods by air-drying the membrane between Whatman filter papers and then storing the membrane, sandwiched between filter papers, in an air-tight container at room temperature or $4^{\circ}C$ (do not store at temperatures below $-20^{\circ}C$ since the membrane will become more brittle, increasing the probability of shattering)

    - to gauge to what extent protein migrated out of the gel, one can stain the gel after transfer

hybridization and washing

---

[2] it's likely that things will work fine without stirring but stirring (1) prevents bubbles from forming under the transfer apparatus and (2) keeps the overall temperature of the transfer apparatus more even

Western blot analysis

1. If the membrane was not incubated in blocking solution overnight at 4°C, immerse the membrane in blocking solution for 15 min to 2 h at room temperature with agitation.

    note: in this lab, the instructor will incubate your membrane in blocking solution overnight at 4°C so that you may proceed to the next step at the start of the lab

2. Decant the blocking solution.

3. Incubate the membrane with primary antibody (diluted in wash buffer) for 20 min at room temperature with agitation.

4. Rinse in 20 mL wash buffer one time.

5. Add an excess of wash buffer to the container holding the membrane and incubate 3 min with agitation.

6. Decant the wash buffer.

7. Incubate the membrane with secondary antibody (diluted in blocking solution) for 15 min with agitation.

8. Rinse in 20 mL wash buffer.

9. Add an excess of wash buffer. Incubate 3 min with agitation.

10. Decant the wash buffer.

visualization

Western blot analysis

1. Using a 10 mL pipet, add 10 mL of HRP color detection reagent to the membrane, allowing it to spread across the surface of the membrane.[3]

2. Incubate for 10 to 30 min with agitation, monitoring color development.

3. When color development is satisfactory, decant the HRP reagent and rinse the membrane twice with distilled water.

4. Blot the membrane dry with a paper towel.

5. Air dry for 30 min to 1 h.

6. Preserve the membrane covered in plastic wrap or under transparent tape on a laboratory notebook page.

Laboratory report

The laboratory report should follow the format described in the first laboratory session and in the course syllabus. This laboratory report should include data from both previous related laboratory sessions and this laboratory session.

Densitometry (see Appendix I, *Densitometry*) might be used to obtain qualitative estimates of protein abundance in samples.

Chapter 12. Reading a scientific paper

analysis of a scientific paper: Gautier *et al.* 1991

The report for this 'lab' does not need to be prepared using a standard lab report format. Simply provide careful, thoughtful answers to the questions below. The questions apply to Gautier *et al.*'s 1991 paper[1] on the biology of cdc25. References below to figures refer to figures in Gautier et al.

1. What considerations motivated these authors to use reticulocyte lysates to generate cdc25 protein?

2. What is the main difference in the experimental procedure employed in Figure 1A and that employed in Figure 1B?

3. In Figure 2A, describe the difference between lanes 2, 3, 4, and 5 in detail. What is the difference between the different forms of cdc25 used in the different lanes? What difference in cdc2 results in the different electrophoretic mobilities observed by SDS-PAGE?

4. What is the conclusion which can be drawn from Figure 2A?

5. Carefully read the text until you understand the experimental procedure used to collect the

---

[1] Gautier J, MJ Solomon, RN Booher, JF Bazan, and MW Kirschner, 1991, cdc25 is a specific tyrosine phosphatase that directly activates p34^{cdc2}, Cell 67, 197

data shown in Figure 3A. Draw a flow-chart representation showing how the experiment was conducted.

Chapter 13. Restriction digest analysis

Restriction digestion of DNA

Introduction

A restriction enzyme (a 'restriction endonuclease') cleaves DNA at a precise sequence. For example, the restriction enzyme EcoRI cleaves double-stranded DNA at the 'recognition sequence'

5'...GAATTC...3'
3'...CTTAAG...5'

to produce

5'...G
3'...CTTAA

and

AATTC...3'
    G...5'

Note that the single-stranded "overhangs" left by this cleavage event are capable of interacting through hydrogen-bonding with other double-stranded DNA molecules with the "complementary" overhangs. Restriction enzymes can be useful both in (1) characterizing a DNA molecule and in (2) cutting a piece of DNA so that it can be inserted into another DNA molecule.

Restriction digest analysis

Figure 13.1. Map of plasmid pTXB1/3

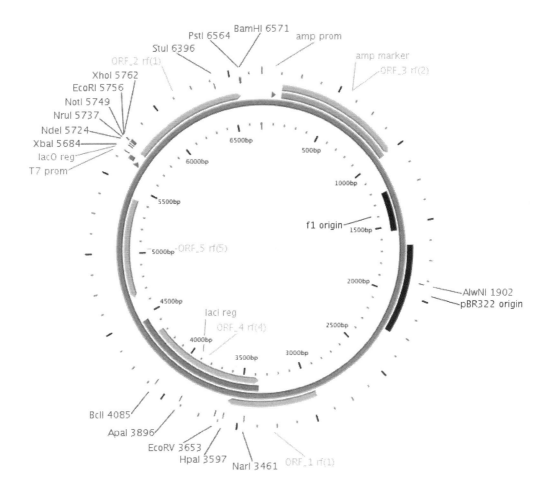

The image was generated with PlasMapper (Xiaoli Dong, Paul Stothard, Ian J. Forsythe, and David S. Wishart "PlasMapper: a web server for drawing and auto-annotating plasmid maps" Nucleic Acids Res. 2004 Jul 1;32(Web Server issue):W660-4)

For example, consider an experiment in which the investigator has isolated a bacterium which the investigator believes was transformed with the plasmid pTXB1. A "restriction map" of a plasmid can be generated by computer (Figure 13.1, "Map of plasmid pTXB1/3") and compared against experimental restriction digest data. In other words, the putative pTXB1 nucleic acid sample can be subjected to treatment with different restriction enzymes. The resulting DNA fragment(s) can then be analyzed by gel electrophoresis and a suitable detection system. The investigator can establish whether the pattern of cleavage events (the number and lengths of the fragments produced with each enzyme) matches the pattern anticipated from the computer-generated restriction map. This data allows the investigator to confirm the identity of the nucleic acid sample. Similar approaches can be used in analysis of gene polymorphisms from human samples for clinical and forensic purposes.

Restriction enzymes are also important when modifying nucleic acid molecules. Plasmids used for expressing genes in various organisms frequently have a "multiple cloning site" (a series of sites cleaved by frequently used restriction enzymes) located directly downstream from a promoter designed to drive expression of the gene of interest in the target organism. If the gene if interest is modified so that it has the appropriate restriction sites adjacent to the coding sequence of the gene, it can readily be inserted at the "multiple cloning site" of the plasmid. From a practical standpoint, this insertion relies on using a restriction enzyme to cut the plasmid to generate "sticky ends", cutting your favorite gene to generate the complementary "sticky ends", placing both DNA molecules in the same tube and allowing the ends to anneal (through hydrogen bonding of the complementary bases), and then covalently

Restriction digest analysis

linking the two annealed molecules using an enzyme termed a ligase.

In this experiment, you will subject a plasmid to restriction digestion with one or more restriction enzymes. You [most likely] do not know the identity of the plasmid. The plasmid may be pGLO, pBLU, pUC19, or pBR322. The sequences for pGLO, pBLU, pUC19, and pBR322 should be available on the course web site (pBR322 and pUC19 can also be accessed via their accession numbers, M77789 and J01749, respectively). Also available online are software tools which can be used to evaluate the theoretical restriction digest products of the plasmids (see "Selected resources" below). Note that sizes and distances of plasmids and plasmid fragments are typically represented in base pairs (bp) or in kilobases (kb or kbp).

The lab report should be a full lab report and your data should include a detailed analysis of the results. For example, your results should include a picture (hand-drawn is fine) of the gel electrophoresis results accompanied by numeric estimates of molecule sizes in the gel electrophoresis portion of the experiment. The theoretical data should include a detailed representation of theoretical cut sites (sequence data and numbering) as well as graphic representations of the theoretical restriction digests.

Procedure

In the first lab session, you will initiate several restriction digests of the sample (a plasmid). In the second lab session, agarose gel electrophoresis will be employed to evaluate the restriction digests.

Restriction digest analysis

In advance: decide which restriction enzymes are appropriate to use. If identifying an unknown plasmid, you should conduct a minimum of two different restriction digests.

1. Plan each restriction digest in advance (see 'planning your digest', below). Each digest should include

    - 1 µg DNA sample

    - 1 µL restriction enzyme

    - 1 µL 10X restriction enzyme buffer

    - 1 µL 10X BSA[1]

    - adjust the final volume to 10 µL with water[2]

2. Add the ingredients of the digest together in a 500 µL snap-cap tube in the following manner:

    - add water first

    - add all other ingredients except for the restriction enzyme

    - mix the ingredients (flick the tube)

    - add the restriction enzyme last: remove it from the -20°C insulated holder and

    - immediately return it to the holder after removing the amount of enzyme you require (restriction enzymes are temperature-sensitive)

Restriction digest analysis

- mix the ingredients; incubate at 37°C for at least 2 h

3. Make sure to save your remaining plasmid DNA!

The steps below will be done the following week.

1. Prepare a gel for gel electrophoresis (see "Electrophoresis" protocol)

2. After incubating, stop the restriction digest:

   - remove the tube from the water bath and add 1.2 µL 10X stop solution

   - mix by flicking the tube repeatedly

   - heat to 65°C for 5 min

   - place the tube on ice for 2 min

3. Perform electrophoresis on the samples of interest. These samples should include:

   - the restriction digest(s) of the unknown plasmid

   - 800 µg of the uncut plasmid (having both the uncut plasmid and the restriction digest on the same gel allows you to confirm that the restriction enzyme was actually active)

   - a series of DNA size standards

   - if you did a "plasmid prep" and didn't obtain sufficient DNA, you should also evaluate this

Restriction digest analysis

sample by electrophoresis to confirm that it really does not contain the expected plasmid DNA

4. On your own: evaluate the restriction digest of the plasmid in comparison to the theoretical restriction digests of the pBLU and pUC19 plasmids.

Laboratory report

The report for this lab should be prepared as a full laboratory report with an Introduction, Materials and Methods, Results, and Discussion section. If electrophoresis is being conducted the following week, complete the report after completing the electrophoresis lab.

Selected resources

A number of software tools are available for sequence retrieval and analysis.

Sequence retrieval

Entrez nucleotide [http://www.ncbi.nlm.nih.gov/entrez/query.fcgi?db=Nucleotide]

An online database of genomic, mRNA, vector, ... sequences from a wide variety of organisms. The database can be searched by accession number (e.g., enter in the search field "Q32310[ACCN]" to search specifically for the gene with accession number Q32310).

Restriction analysis resources

NEBCutter [http://tools.neb.com/NEBcutter/index.php3]

Restriction digest analysis

This online tool evaluates restriction enzyme sites in a sequence of interest. Nice graphical representations, etc. Remember to indicate if the sequence is circular or linear. Note: NEBCutter does not print hard-copies correctly under Windows 98 and Netscape 6.1.

Try the "custom digest" option.

WebCutter [http://www.firstmarket.com/cutter/cut2.html]

This online tool evaluates restriction enzyme sites in a sequence of interest. Remember to indicate if the sequence is circular or linear (select "Circular Sequence Analysis"). Version 2.0 does not work well with Netscape 6.1.

New England Biolabs [http://www.neb.com]

The restriction enzyme company. Online catalogs, technical resources, vector sequences and maps, etc.

ReBase [http://rebase.neb.com/rebase/rebase.html]

This online database of restriction enzyme characteristics contains links to a number of tools for identification of restriction sites in your molecule of interest. REBsites [http://tools.neb.com/REBsites/index.php3] is an online tool which gives convenient graphical representations of restriction digests of a sequence of interest. Remember to indicate if the sequence is circular or linear.

Bioedit [http://www.mbio.ncsu.edu/BioEdit/bioedit.html]

Restriction digest analysis

This Windows software can be used to evaluate restriction enzyme sites in a sequence of interest. Moderately user-friendly. Crashes under Windows 98 when run in combination with certain other software packages.

GeneDoc [http://www.psc.edu/biomed/genedoc/]

This Windows software can be used to evaluate restriction digest patterns. Early versions were not very user-friendly nor intuitive to use.

Labvelocity [http://www.labvelocity.com]

This site has vector sequences and maps, among a host of other useful tools.

Chapter 14. The PCR

Introduction

Review the section on PCR in the course text.

Detecting minute amounts of DNA is a common application of PCR (PCR can really be used to get evidence from 'no clues left behind' crime scenes - a single drop of blood or a hair can be sufficient to generate data regarding the identity of the corresponding individual).

Procedure

overview

week 1: start the PCR reaction

week 2: electrophoresis of PCR reaction products; staining of gel to visualize results

week 1: prepare and run the PCR reaction

1. Ensure that the "crime scene" DNA sample, "suspect DNA" samples, and primers are thawed. Store the tubes on ice during the experiment.

2. Obtain aerosol barrier pipet tips (PCR is sensitive enough that even a trace amount of contamination may alter experimental results).

3. Obtain PCR tubes for each of the samples. Label the tubes you will use in the PCR reaction as follows:

The PCR

1 tube: "CS" (crime scene reaction)

1 tube: "A" (suspect A reaction)

1 tube: "B" (suspect B reaction)

1 tube: "C" (suspect C reaction)

1 tube: "D" (suspect D reaction)

4. You will be given a tube, labeled "MMP" (master mix plus primers), containing 120 µL of the 'master mix' (Taq polymerase, dNTPs, and $MgCl_2$ buffered at pH 8.0) and the primers for the reaction.

5. Add 20 µL of MMP solution to the 'CS' tube.

6. Return the tubes to the ice bath (use a foam float) to hold the tubes.

7. Repeat the above steps for the A, B, C, and D tubes.

8. **Carefully** remove 20 µL of Suspect A DNA solution from the 'Suspect A' tube located at the front of the class and transfer the DNA solution to your "A" tube.

9. Return the tubes to the ice bath (use a foam float) to hold the tubes.

10. Repeat the above steps for the 'Suspect B', 'Suspect C', 'Suspect D', and 'Crime Scene' samples. **Ensure you use a fresh pipet tip for each sample.**

11. Ensure the lid of each PCR tube is tightly shut.

The PCR

12. Place the PCR tubes in the thermal cycler.

13. The instructor will run the PCR program as follows:

    initial denature (1 time): 94°C, 2 min

    thermal cycling (35 times):

    94°C, 30 sec

    52°C, 30 sec

    72°C, 1 min

    final extension (1 time): 72°C, 10 min

    hold (indefinite): 4°C

Evaluate the PCR reaction using agarose gel electrophoresis.

Chapter 15. Agarose gel electrophoresis of DNA

Introduction

Gel electrophoresis is widely used as a means to separate mixtures of nucleic acids and to separate mixtures of proteins. Gel electrophoresis is employed both for analytical work -- e.g., for estimating the size and/or amount of a protein or nucleic acid -- and as a preparative tool. PAGE can be used as a matrix for separating smaller oligonucleotides (tens to hundreds of nucleotides in length). Agarose is generally used for larger nucleic acids (hundreds to thousands of nucleotides in length). A variety of stains are available for nucleic acid detection. Ethidium bromide is a fluorescent molecule which specifically binds nucleic acids and, until the late 1990s, was the predominant stain used for standard agarose gel electrophoresis experiments. Newer fluorescent stains such as Sybr Gold, Sybr Green I, Sybr Green II, and Sybr Safe, are now often used in place of ethidium bromide since they have either higher sensitivity or reduced mutagenicity. Size standards are typically used to estimate the sizes of DNA molecules in a sample. A frequently used size standard is lambda phage DNA digested with the HindIII restriction enzyme. This document describes a protocol for evaluating a double-stranded DNA sample by agarose gel electrophoresis.

Agarose gel electrophoresis of DNA

Figure 15.1. Visualization of apoptotic DNA

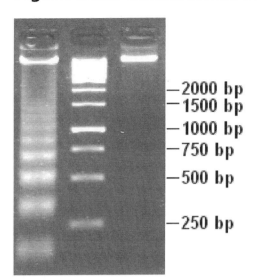

Lane 1 contains DNA extracted from apoptotic cells. Lane 2 contains a DNA mass ladder. Lane 3 contains DNA from normal cells

general notes:

- to participate in this lab, **you must prepare a sample loading table and complete as much of it as possible before the start of the lab** (see the protocol below)

- avoid exposing your skin or eyes directly to ultraviolet light (use gloves and goggles)

- for the well-prepared student this lab should take no more than 2.5 to 3 h

Procedure

Wash the gel holder and comb (the comb you use should have less than ten teeth/wells)

Figure 15.2. Disassembled minigel apparatus including wedges, comb, and tray

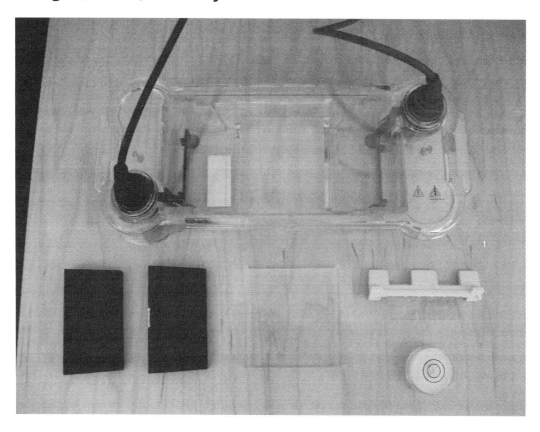

Figure 15.3. Tray in minigel apparatus

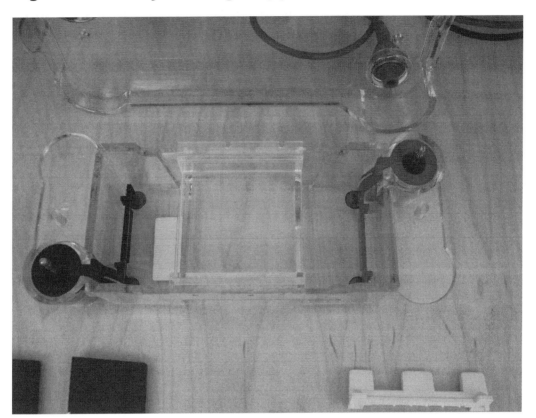

Pour the electrophoresis gel:

1. Clean the gel electrophoresis "box" and accessories (e.g., wedges, comb(s), etc.).

2. Seal the ends of the gel holder (using wedges or autoclave tape).

    - if using wedges, place the tray in the gel electrophoresis box (Figure 15.3, "Tray in minigel apparatus"), then place the wedges at each end of the tray (Figure 15.4, "Tray with wedges in minigel apparatus")

Agarose gel electrophoresis of DNA

note: ensure the ends are really sealed (it's not too exciting to prepare another agarose solution after the first solution leaks out)

Figure 15.4. Tray with wedges in minigel apparatus

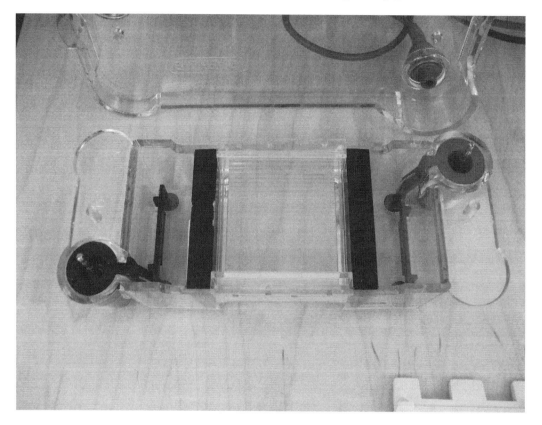

1. Prepare the electrophoresis buffer (1X SB unless the instructor indicates otherwise) from the concentrated stock if necessary (450 mL will be required for preparing the gel and to fill the chambers adjacent to the gel).

2. Prepare the agarose. For analyzing restriction digests, prepare a 0.8% gel (0.32 g agarose and 40 mL of electrophoresis buffer). For analyzing

Agarose gel electrophoresis of DNA

restriction digests and PCR products, prepare a 1.6% gel (0.64 g agarose and 40 mL of electrophoresis buffer). For analyzing only PCR products, prepare a 2.5% gel (1 g agarose and 40 mL of electrophoresis buffer).

hint:[use a 200 mL Erlenmeyer flask for preparing the solution]

3. Boil or microwave the suspension until a uniform solution is formed (note: be careful to mix regularly if using the microwave to avoid superheating; you can invert a small Erlenmeyer flask in the mouth of the larger flask to 'reflux' and reduce loss of volume while boiling).

4. Allow the solution to cool until the solution isn't steaming.

5. **Before** the solution begins to solidify, pour the agarose solution into the tray. **Immediately** center the comb over the notches corresponding to the end closest to the (-) terminal; pop (or drag to the side) any bubbles in the gel, if present, using a yellow pipet tip.

6. Allow the gel to polymerize on a level surface (appr. 0.5 h at room temperature).

prepare samples and standards

A loading table, describing the samples and standards, should be prepared in advance.

Preparing a sample from a restriction digest:

Agarose gel electrophoresis of DNA

- add 1 µL of 10X STOP buffer[1] to a 10 µL restriction digest

- heat at 65°C for 5 min

- incubate on ice 2 min before loading

- leave at room temperature until ready to load

notes for restriction digests:

- Evaluate both the 'cut' sample(s) and the original, uncut, sample DNA. This allows you to evaluate whether a restriction enzyme has lost activity.

  - For restriction digests and plasmids which have not been previously evaluated by electrophoresis, evaluate at least 2 µg (if possible) by electrophoresis.

Preparing a sample of uncut plasmid:

- load 1 to 2 µg of plasmid

- do not throw away the remainder of your uncut plasmid sample; return the tube to the -20°C freezer

Preparing a molecular mass standard:

- use the instructions provided by the instructor

  example: for the Fermentas FastRuler DNA Ladder (middle range), the ladder is premixed with loading buffer; load 20 µL per lane

Preparing a PCR product:

Agarose gel electrophoresis of DNA

- spin the tube to collect liquid at the bottom of the tube

- add 10 µL Orange G loading dye to a 10 µL PCR reaction

  note: only <u>you</u> know your reaction volume (hint: consider checking your notebook)

- load all of the sample (20 µL)

Table 15.1. Example of planning for loading if one was evaluating sensitivity with an uncut plasmid

Lane	1	2	3	4
Sample	molecular mass standard	uncut plasmid	uncut plasmid	uncut plasmid
Concentration ($\mu g/\mu L$)		0.5	0.5	0.5
Sample mass		50 ng	500 ng	2 μg
Volume (μL)				
Sample		0.1	1	4
10X STOP buffer		1	1	1
Electrophoresis buffer		8.9	8	5
Total volume	10	10	10	10

gel electrophoresis

1. Once the gel is polymerized,

   - remove wedges or autoclave tape

   - remove (carefully!) the comb

   - ensure the gel is oriented so the lanes run from (-) to (+)

2. Add the electrophoresis buffer to the chambers adjacent to the gel until the buffer level is just

Agarose gel electrophoresis of DNA

above the top of the gel and has filled the wells (rock the gel apparatus gently back and forth if the buffer does not immediately fill the wells).

3. Position the gel electrophoresis unit so that the lid, when plugged into the power supply, can be placed on the electrophoresis unit.

4. Load the gel:

    - draw the sample into pipet tip (ensure there are no bubbles in the tip/sample), place tip directly above well, with the tip immersed in electrophoresis buffer, slowly depress pipettor plunger until all sample has settled in well

    hint: avoid overflowing the well; if this happens, the sample may end up in other lanes of the gel, generating results that are more difficult to interpret

5. Run the gel:

    - 1XSB: 6-35 V/cm (e.g., 180V for a 7 cm gel works fine; running at 240 V will melt the gel)

    - NOTE: for the Biorad 'PowerPac300', you must press the 'runner' button to actually subject the gel to the desired voltage

6. Stop the gel when the bromphenol blue dye front is less than 0.5 cm from the end of the gel.

7. Remove the gel, carefully physically supporting the gel to prevent damaging the gel. Place the gel in a plastic tray.

Agarose gel electrophoresis of DNA

8. Stain the gel with Sybr Safe:

    - add 60 mL of electrophoresis buffer to the tray

    - add 6 µL of 10000X Sybr Safe (see instructor to obtain Sybr Safe)

    - incubate in the dark with gentle agitation for 20 to 30 min

9. Convey the gel to the UV illuminator and evaluate the gel using 254 nm illumination (a metal spatula seems to provide a good background).

10. Illuminate with UV light and document your observations.

Laboratory report

The laboratory report associated with this experiment includes data from one ore more preceding laboratory sessions. You should also integrate this information into your lab report.

To begin to analyze your data, consider these questions:

What were the apparent sizes of the DNA electrophoresed in each lane?

How do the observed data compare with the theoretical data?

Chapter 16. Physarum

Introduction

A collection of *Physarum polycephalum* cells can behave as an organized unit, exhibiting collective behaviors such as phototaxis and chemotaxis. Rather than using a motion system like that of muscle cells, the mobility of the *Physarum* cell collection is realized via coordinated cytoplasmic streaming.

Procedure

observe cytoplasmic streaming

- Obtain an established culture of *Physarum polycephalum* (it should look like a clear yellow layer on agar covering at least a roughly dime-sized region of the medium)

- Observe under 40X total magnification

  - if stage has slide clip, ask instructor if it is possible to remove the clip (typically such clips are held in place with 'thumb' screws) - removing the clip will make it easier to manipulate large plates on the stage

  - place the entire culture plate on stage

- Record your observations of the *Physarum*

  - do there seem to be different morphologies the *Physarum* adopts in different regions of the dish? different tube types?

- where does cytoplasmic streaming seem to be occurring most?

- estimate the average rate of cytoplasmic streaming

  - is cytoplasmic streaming unidirectional?

is a *Physarum* 'organism' phototactic?

- Obtain an established culture of *Physarum polycephalum*

- Obtain a 100 mm Petri dish containing 2% agar or 2% oatmeal agar

  - use a black pen to draw a line describing the diameter of the dish

  - mark one half of the circle "L" and the other half of the circle "D"

- Excise an approximately 0.5 cm x 0.5 cm piece of agar (with plasmodium on the surface of the agar) from the dish and transfer the agar to the center of the new plate

  - ensure the agar block is placed plasmodium-side-down on the new dish

- If the dish contained 2% agar, scatter a few oats on each side of the dish, ensuring that on each side at least one oat is within 1 cm of the block

- Cover the plate with aluminum foil

- Cut an approximately 1 cm^2 piece of aluminum foil approximately 1 cm from the edge on the "L" side of the dish

- Place the dish in a location where it will remain at approximately room temperature and will be illuminated at least during day light hours

- Record observations at approximately 24 and 48 h

testing a hypothesis regarding chemotaxis

In advance, select at least two compounds or solutions to test for chemotactic potential with *Physarum* from the list provided by the course instructor.

note: multiple plates can be used to evaluate a wider variety of chemicals

Prepare samples:

- For liquid samples,

  - prepare each test solution in a 50 mL tube

  - place a small (appr. 4 mm diameter) piece of filter paper in the tube and allow the solution to saturate the paper

  - remove the paper from the tube with tweezers; allow excess liquid to drip off of paper; place the paper on agar with the closest point approximately 0.5 to 1 cm from the agar block with plasmodium

- For solid samples,

- cut sample into pieces approximately 0.5 cm x 0.5 cm x 1-2 mm (W X L X H)
- place several pieces on the plate with the closest piece less than 1 cm from the agar block with the plasmodium

Prepare the culture:

- Obtain an established culture of *Physarum polycephalum* (clear yellow layer on agar covering at least an approximately 2 cm2 region of the medium)
- Obtain a 100 mm Petri dish containing 2% agar or 2% oatmeal agar
  - use a black pen to draw lines on the bottom of the dish, dividing the dish into quadrants
- Excise an approximately 0.5 cm by 0.5 cm piece of agar (with plasmodium on the surface of the agar) from the dish and transfer the agar to the center of the new plate
  - ensure the agar block is placed plasmodium-side-down on the new dish
- Place one sample in each quadrant; as a positive control, one can scatter a few oat flakes (ensure at least one flake is within 1 cm of the agar block) in one of the quadrants
- Incubate in the dark at approximately room temperature [optional: use humidified chamber]
- Record observations at appr. 24 and appr. 48 h.

experiment: what happens to the original culture if it isn't fed?

- Remove the *Physarum* culture dish from the humidifying chamber (if using a humidifying chamber)

- Open the lid of the dish slightly to permit moisture to exit

- Don't add additional oat flakes

- Observe the culture at 4 d, 1 week, and 2 weeks, noting changes in appearance/morphology and cytoplasmic streaming

Appendix A. Sample lab safety handout

Etiquette

If you did not purchase it, were not given it, and you did not earn it, it probably does not belong to you.

Please do not remove laboratory items from the lab area without the instructor's permission.

Do not use/touch/move a sample that belongs to someone else (i.e., if a sample tube/vial/etc. does not have your initials on it).

Glassware and plasticware

Unless otherwise instructed by the course instructor, clean plasticware and glassware after use. The method used to clean labware depends on the contaminant.

Hydrophobic (e.g. greases): Rinse with a dilute detergent solution (< 2% strength) or ethanol/isopropanol followed by thorough water rinsing (avoid stronger solvents such as acetone or methanol with plasticware).

Biological: Soak in a 1:10 dilution of bleach or use wescodyne or 70% ethanol to disinfect; rinse extensively with water.

Hydrophilic (e.g., salts, buffers, etc.): Extensively rinse with water.

After cleaning, invert on a paper towel to dry.

Safety

Don't rush -- relax -- mistakes happen most easily when you are uptight or hurrying

In the case of an emergency, notify the instructor at once.

Note the locations of the

- emergency gas shutoff

- first-aid kit(s)

- fire extinguisher(s), fire alarm(s), and fire blankets

- eye-washes

- showers

Figure A.1. A safety shower and eyewash

Think out in advance how to deal with and prevent fires. Very small fires on the bench can be allowed to burn or smothered with wet towels. Dispose of flammable liquids in waste bottles in the hood. If a substance emits flammable vapors, keep it in the hood at all times -- especially if burners are being used in the laboratory.

Pregnancy: if you are pregnant or planning on becoming pregnant, it is worth considering taking the class at another time since there is potential risk of exposure to teratogenic compounds.

The bench

Do not store "non-laboratory" items (e.g., your coat or textbooks) on the bench -- the goal is to keep chemical or microbiological 'nasties' from migrating out of the lab with you

Clean the bench after use. Use the same method as would be appropriate for cleaning plasticware or glassware.

Using the hood

1. Make sure the hood is functional.

2. The hood sash should be closed when not in use.

3. The sash position where safe flow exists should be marked on the side of the hood. Closing the sash further increases the hood's effectiveness but may impair freedom of movement. It is generally a good idea when working with very volatile compounds to close the sash to the greatest extent possible which still permits freedom of movement in conducting the manipulation. However, exercise care when working with powders or flames since lowering the sash can produce stronger air currents in the hood.

Apparel

Consider wearing safety glasses whenever you are doing laboratory work. In particular, wear safety glasses or goggles if there is a chance of splashing or an aerosol being formed -- also wear safety glass during manipulation of vials during extreme temperature transitions. Contact lenses are NOT allowed.

Tie hair back -- the goal is to keep your hair from wandering around in solutions, wrapping itself around lab equipment, or igniting in bunsen burners.

Wear your lab coat -- this protects you from spills.

Wear close-toed shoes (i.e., no flip-flops, sandals, ...) -- this protects you from spills.

In the case of a hazardous chemical spill, unless otherwise indicated, it is safest to immediately remove the clothing affected -- if the chemical has reached the skin or is in the eye, rinse with copious amounts of water (locate the eyewashes in the laboratory).

Always wear gloves -- make sure to choose the appropriate type of glove for the work you are doing. Latex gloves are appropriate for use with microbiologicals and water-soluble compounds. Nitrile gloves should be used when working with organic solvent-based solutions.

Personal hygiene

Wash hands and dry them after concluding an experiment and before leaving the laboratory -- you probably don't want to eat chemical residues along with the food, gum, etc. you touch after leaving the lab.

Don't apply cosmetics or insert contact lenses in the laboratory.

Sample lab safety handout

Avoid eating, drinking, or tasting anything while in the laboratory. Avoid smelling lab reagents.

Accidents

Report accidental cuts/scrapes, burns, or chemical spills to the instructor immediately.

Liquid and solid manipulation

Information about laboratory hazards associated with a reagent is available in the form of a "Material Safety Data Sheet" (MSDS). The student should be familiar with the hazards presented by each compound which will be used in a given laboratory session. One means of acquainting oneself with the hazards presented by a compound is to review the MSDS for that compound. MSDS's are available online at several locations (e.g., http://www.sigma-aldrich.com or http://www.setonresourcecenter.com).

Do not insert anything in a reagent bottle -- including a "clean" spatula or dropper. Do not return excess material to a class reagent bottle. If a stopper or lid seems stuck, see the instructor.

Never mouth pipette -- only use mechanical pipeting aids.

For powders which readily suspend (e.g., SDS), always use a face mask while measuring or pouring the powder.

If a substance produces fumes, it should be handled in a fume hood.

Spills: do not clean a chemical spill unless you know the proper means of neutralizing the spilled substance. If in doubt or unsure, notify the instructor. Acids and bases should be neutralized prior to any cleanup effort. Most other compounds should be adsorbed with activated charcoal.

Always add acid to water, never water to acid.

Working with flames

Don't -- unauthorized flame use is not permitted in the lab. Flames present an extraordinary danger when used in the proximity of organic solvents, many of which are highly flammable.

Appendix B. Statistics software

Online statistics resources

In the past, Interactive Statistical Calculation Pages [http://statpages.org/] has featured a relatively comprehensive list of web pages that perform statistical calculations. Such pages range from relatively powerful tools such as Rweb [http://www.math.montana.edu/Rweb/], a web-based interface to the R statistical package, to SISA [http://www.quantitativeskills.com/sisa/index.htm], a straightforward web-based interface to tests such as the T-test and calculations such as Bonferroni correction, and Simple linear regression page [http://people.hofstra.edu/faculty/Stefan_Waner/newgraph/regressionframes.html], a web-based interface for performing linear regression.

Statistics software

Most spreadsheet software (e.g., OpenOffice, KOffice, gnumeric) includes quite a bit of statistics functionality. Take the time to read the documentation for the statistics functionality of the package.

Some software with statistics support

gfit [http://gfit.sourceforge.net/]

- model-based global regression

gnumeric [http://projects.gnome.org/gnumeric/]

gnuplot [http://www.gnuplot.info]

Statistics software

- linear regression and more sophisticated fits

LibreOffice [http://www.libreoffice.org/]

MODELbuilder [http://www.modelbuilder.org]

- model multivariate data

PSPP [http://www.gnu.org/software/pspp/pspp.html]

- aims to replace functionality of SPSS statistics package (including descriptive statistics, T-tests, linear regression and non-parametric tests)

R [http://www.r-project.org/]

- a statistics package with a lot of power but also with a substantial learning curve; consider using one of the R GUIs to get started

statist [http://statist.wald.intevation.org/]

- console statistics program w/interactive menu
- uses gnuplot for graphics output

tol [http://www.tol-project.org/]

- focused on time series analysis and stochastic processes

Xlisp-Stat [http://www.xlispstat.org/] - a statistics package which can perform regression and other types of analyses on a data set

- see also Vista, a free program, based on Xlisp-Stat, for analyzing data and generating statistics and plots

Appendix C. Mass measurement

Mass measurement

temperature effects

Unless temperature equilibrium is established, the apparent mass of the object will fluctuate.

- allow sample temperature to adjust to the ambient temperature before weighing

- use tweezers or another method to indirectly manipulate the object (using one's hand, even if gloved, is likely to transfer heat to the sample)

electrostatic charge

Certain materials can accumulate static charge, typically due to friction. As the charge dissipates, the apparent mass will fluctuate.

- use an anti-static tool to neutralize surfaces

- use anti-static products (gloves, weigh boats, ...)

- maintain air humidity

References

"Sigma-Aldrich, 2007, Weight Measurement, the Accurate Way." [http://www.sigmaaldrich.com/etc/medialib/flashapps/labware-notes/pdfs/labwarenotes-v1-3.Par.0001.File.tmp/labwarenotes_v1_3.pdf]

Appendix D. Microscopy

Introduction

The light microscope is heavily used in biology. It is an invaluable tool for the cell or molecular biologist as well as for the clinician. The basic 'building blocks' of higher organisms, **cells**, are too small to be seen with the naked eye. The light microscope allows us to visualize cells and, in the case of eukaryotic cells, some aspects of the internal structure of cells. In cell biology, the ability to evaluate samples visually at the microscopic level is essential. Microscopy is regularly employed to assess the health of cell cultures and to evaluate cell morphology. It also can be a useful tool in more specialized procedures such as monitoring the progress of a subcellular fractionation or evaluating cellular conversion of a substrate to a colored product. A specialized form of microscopy, fluorescence microscopy, has enriched science immensely over the past decades with the wide availability of probes for diverse applications[1].

With light microscopy, samples can be prepared as 'wet mounts' or as fixed specimens. A 'wet mount' is a sample which is immersed in an aqueous solution and then visualized. One strength of this technique is that it facilitates the observation of living cells. A disadvantage of a wet mount is that the sample cannot be preserved in such a fashion indefinitely; any observations must be made in a relatively short time period following preparation of the specimen. In

---
[1] If curious, see the Molecular Probes web site.

contrast, 'fixing' a sample (preserving the sample using chemical and/or physical methods) allows the sample to be retained for observation for months to years.

Visualization of samples can be enhanced by **staining**. For samples which are relatively transparent, it can be difficult to distinguish structural details using light microscopy. Staining is a process where a live or fixed sample is treated with a compound which specifically binds, or is excluded from, specific regions of the cell. The compound typically is colored, although these days fluorescent and infrared dyes are often employed in staining. Staining enhances visualization by increasing the contrast between different regions of the cell and/or between the cell and the surrounding environment.

Guidelines for microscope use

- Exercise care in handling the microscope. When moving it from one place to another, the microscope should be gripped tightly with both hands and never held gripping the stage or the optics.

- Only clean the lenses with lens paper, never with a Kimwipe or other paper -- realize that the objectives are by far the most expensive components on the microscope.

- Take great care in adjusting the stage height and when switching from a low objective to a higher objective lest the lens hit the slide.

- Always raise the objective before placing a slide or a hemocytometer on the stage or removing a slide or a hemocytometer from the stage

Familiarizing yourself with the compound microscope

Figure D.1. A compound microscope

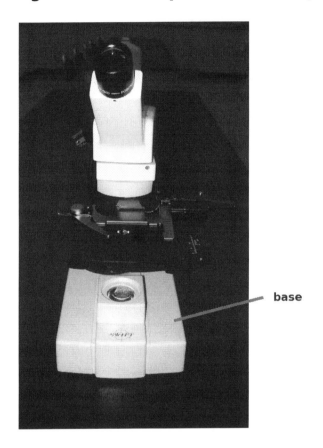

The compound microscope (Figure D.1, "A compound microscope") is a microscope which uses multiple lenses to magnify the image of the sample. The compound microscope is expensive and relatively fragile. Ensure that you never move the microscope unless you are supporting it by its base or its arm (Figure D.2, "The microscope arm and stage controls").

Figure D.2. The microscope arm and stage controls

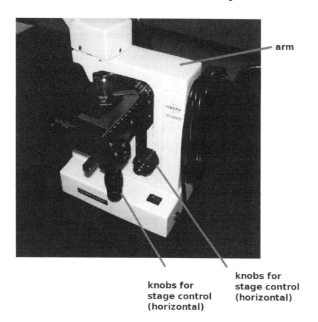

Obtaining a 'good image' using the compound light microscope can take a bit of effort. An understading of a few basic optics principles and of the roles of the microscope components can aid the user in optimizing image quality.

the objectives

The microscope objective (Figure D.3, "The microscope objective") is a central component of the microscope. The objective serves to gather light from the specimen and magnify the image. Objectives are expensive: treat them with care.

Figure D.3. The microscope objective

On the side of the objective barrel, a set of labels describe various qualities of the object. The <u>magnification</u> (e.g., "4X" or "40X") describes the extent to which the image is magnified by the objective. The <u>numerical aperture</u> (a value which can range from 0.1 (low-magnification lenses) to 1.6 (specialized high-magnification lenses)) represents the resolving power of an objective (think of answering the question, "what is the minimum physical distance between two points at which those points can be recognized as distinct points?").

A second objective is the eyepiece objective. This is often a 10X objective. If one were viewing a specimen using a 40X objective with a microscope with a 10X eyepiece, the total magnification would be the product of the magnification values for the individual objectives, 400X.

the stage

Figure D.4. The microscope stage

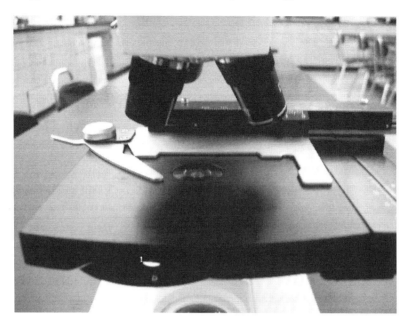

The stage (Figure D.4, "The microscope stage") is the platform on which the sample is placed. Typically, a spring-loaded arm on the stage holds the Figure D.5, "A slide on the microscope stage" in a fixed position. The stage can be moved using knobs (Figure D.2, "The microscope arm and stage controls") which control motion in the two dimensions defined by the stage.

Figure D.5. A slide on the microscope stage

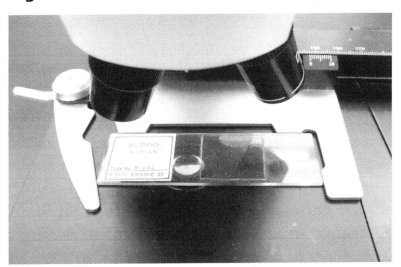

Depending on the instrument, the stage can also be moved in the vertical plane or the objectives can be moved in the vertical plane using the coarse and fine focus adjustment knobs (Figure D.2, "The microscope arm and stage controls"). The end result in either case is to alter the distance between the objective lens and the specimen. Altering the distance between the objective and the specimen is used to bring the specimen into focus. The adjustment should be done with care so that the specimen does not 'crash' into the objective.

the light source

The light source (Figure D.6, "The microscope light source") is typically located at the base of the instrument. There is usually a control near the light source for varying the intensity of light. When first viewing a specimen, turn this control down and then raise the light intensity to the minimum amount required to visualize the specimen. Typically, higher intensities of light will "wash out" the specimen, often to a point where it is no longer visible.

Figure D.6. The microscope light source

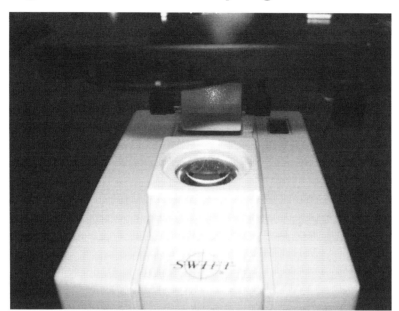

the condenser

The condenser is the structure which adjusts the light *before* it passes through the sample. Adjusting and aligning the illumination is a critical aspect for obtaining high quality images. Although we are not using $10,000 microscopes in this exercise, it is still worthwhile to consider the basics of adjusting illumination. Typically, the vertical position of the condenser and the size of the condenser aperture diaphragm are used to adjust the 'cone' of light that illuminates the sample. As the condenser diaphragm is closed, the amount of light reaching the specimen lessens and the sample becomes 'dim'. This type of adjustment can also reduce undesirable glare and minimize the effects of any imperfections (e.g., dust) on the optical surfaces of the condenser optics. On some microscopes, the condenser is not very sophisticated and does not have an adjustable vertical position nor a continuously variable diaphragm (Figure D.7, "A simple condenser").

Figure D.7. A simple condenser

Adjusting the condenser (if the microscope has an iris diaphragm[2]):

- generally, the substage condenser should be adjusted as close to the stage as conditions permit

- focus with the 4X objective

- switch to the 10X object

- focus with the 10X object

---
[2]sometimes the iris diaphragm is referred to as the field diaphragm

- adjust the iris diaphragm :

  - close the field diaphragm almost completely

  - focus the spot of light by slightly lowering the condenser height

  - if the spot of light is not centered, use the centering screws which protrude from the condenser assembly to center the condenser

  - continue opening the field diaphragm until the size of the spot of light is approximately that of the viewing field

  - note that the bottom line is image quality: if better image quality is obtained by closing the field diaphragm slightly more, then it makes sense to do so...

- if using a microscope without a field diaphragm, pull out an eyepiece and focus and center the spot of light by looking down the eyepiece tube following the above steps

the interpupillary distance

Finish preparing the microscope for use by adjusting the interpupillary distance (the distance between the eye tubes):

- bring the specimen into focus using the coarse focus knob followed by the fine focus knob

- then adjust the distance between the eye tubes, typically by applying gentle pressure to the side of the structure on which the tubes are mounted, to coincide with your interpupillary distance

Additional steps you may want to perform (these may yield better results)

- adjust the diopter focus: turn eye lenses (usu. clockwise) to the shortest focal length position, bring the 10X objective into the light path and focus the specimen; rotate to a lower magnification (e.g., 4X or 5X); refocus each eye lens individually until the specimen is in sharp focus; rotate the 20X objective into the optical path; focus with the fine focus; repeat the diopter eye lens adjustments with the lower magnification objective (e.g., 4X or 5X)

- whenever changing to a different objective, adjust the condenser and field diaphragm as described above

- adjust/center the lamp bulb filament[3]

  - if the field of view is unevenly illuminated, the bulb filament is probably not centered

Cleanup

All slides should be disposed of in the appropriate containers:

- biohazard material: anything which has come in contact with a bodily fluid

- non-biohazard slides and cover slips can be disposed of in the broken glass box

Microscopy

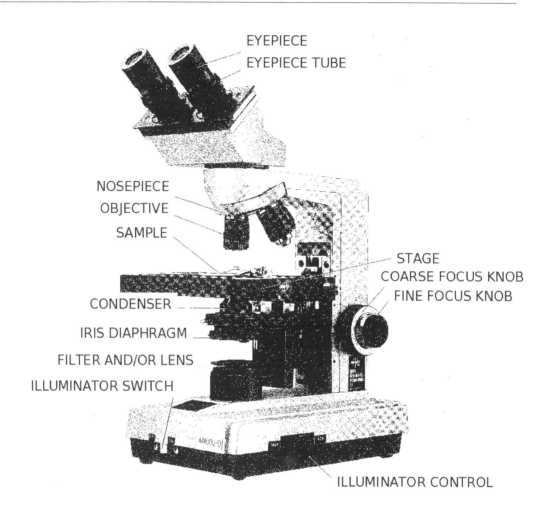

Appendix E. Sterile technique

Bacterial culture: sterile technique

To culture microorganisms or mammalian cells, a technique must be used which minimizes the probability of introducing unwanted organisms into the culture. Sterile is defined as the absence of life. There are varying degrees of sterile technique, applicable to settings which require different degrees of sterility. For bacterial work, sterility is typically less stringent than for mammalian cell culture.

Some guidelines for bacterial sterility ("bacti sterile"):

1. All glassware and plasticware which comes in contact with the culture should be autoclaved to ensure sterility. Autoclaved glassware can be autoclaved with aluminum foil over any openings and the aluminum foil removed directly prior to use of the glassware. If the external surface of the glassware will come in contact with the culture, the whole glassware item should be wrapped in aluminum foil before autoclaving. Autoclaved plasticware should be stored in closed glass or plastic containers or wrapped in a manner similar to that described for glassware.

2. When autoclaving, autoclave tape should be applied to the container to ensure that the autoclave cycle has completed properly (the tape doesn't guarantee it completed properly but it at least indicates that something happened...).

Sterile technique

3. Gloves should be worn during all manipulations -- frequently rinsing the gloves with 70% ethanol will disinfect them and reduce the chances of introducing microorganisms into the material being manipulated.

4. The benchtop should be thoroughly cleaned and wiped with 70% ethanol before manipulations are performed on the bench.

5. Sterile pipet tips should be used for all manipulations of small volumes of liquids. Common sense dictates that the pipet container box is closed immediately after a pipet tip is removed from it.

6. Solids should be dissolved in solution and either filter-sterilized (into a sterile container) or autoclaved (depending on the thermolability of the solid) before use in bacterial culture.

7. All liquids used for culture should be autoclaved. Make sure that the lids are loose during the autoclave process and that glass containers containing liquids are resting in a pool of water. Also ensure that the autoclave is set to a cycle that does not rapidly vent at the end of the pressurization phase.

8. Avoid having openings to containers in a plane parallel to the ground for periods any longer than are absolutely necessary (i.e., try to avoid having stuff fall in the mouth of the container).

9. To sterilize loops or glass spreaders, immerse the region which will contact the culture in 70%

ethanol. Remove from the ethanol solution and flame the wet region. If necessary, cool the spreader by touching it to a region of agar prior to spreading (otherwise, it may be hot enough to kill the microorganisms you are interested in culturing).

Appendix F. Screenshots

Printing a screenshot

Windows

Use the "Print Scrn" key to copy the screen to a buffer. Run the "Paint" program and paste the buffer into the paint window. You may want to use the "Invert colors" tool or other processing to make the image more suitable for printing if it originally had a black background. Then, go ahead and print the image.

Linux

Run the gimp program (type "gimp" at the command line). Make sure the window you want to take a screen shot of isn't blocked by other windows. In the gimp window, click File, then select Acquire, then select Screen shot. Ensure the checkbox next to "single window" is selected. Click the Grab button. The mouse pointer will change to a crosshairs pattern. Place the crosshairs pattern on the window you want to capture. Click the mouse button. Right click on the image to bring up a menu which allows you to print and/or save the image.

Appendix G. Protein assays

Protein concentration assays

Since proteins are the workhorses of the cell, scientists are frequently interested in the relative abundance, activity, or biochemical characteristics of a protein in a sample. Samples may consist of relatively purified protein from a protein purification protocol or may consist of crude cell or tissue homogenates ("lysates"). In all cases, it is important to determine the amount and concentration of protein in the sample before subjecting it to further manipulations. There are several means of estimating protein concentration in a sample, each with its pros and cons; several of the more commonly utilized techniques are briefly described below.

Spectrophotometry

Absorbance assays depending on a protein's intrinsic UV absorbance (typically at 280 nm and 205 nm) are occasionally used to evaluate protein concentration. These assays are excellent for use with relatively pure protein solutions; however, the solution must be free of other UV-absorbing substances (e.g., absorbance at 205 and 280 nm is subject to interference from nucleic acid or lipid components in a sample).

Bradford assay

The Bradford total protein quantitation assay (Bradford, Anal. Biochem. 72: 248, 1976), and several commercial modifications thereof, are colorimetric assays based on the tendency of the dye Coomassie G-250 to shift absorbance from 465 nm to 595 nm (there is a simultaneous color change of the reagent from red/brown to blue) when the reagent binds proteins in an acidic solution. The mechanism of the reaction is based on an anionic form of the dye; this form interacts primarily with arginine residues and, more weakly, with histidine, lysine, tyrosine, tryptophan, and phenylalanine residues. The reaction reaches a relatively stable endpoint so valid absorption measurements can be made over the course of the hour following equilibration of the dye and protein.

The Bradford assay is subject to interference from some compounds (see the Bradford protocol).

In any protein assay the best protein to use as a standard is a purified preparation of the protein being assayed. In the absence of such an absolute reference protein another protein must be selected as a relative standard. The best relative standard to use is one which gives a color yield similar to that of the protein being assayed. Selecting such a protein standard is generally done empirically. Alternatively, if only relative protein values are desired, any purified protein may be selected as a standard. If a direct comparison of two different protein assays is being performed, the same standard should be used for both procedures. With the Bradford protein assay, the dye color development is significantly greater with albumin than with most other proteins, including gamma-globulin. Therefore, although an albumin standard is commonly used, for a color response that is typical of many proteins, the gamma-globulin standard is appropriate. (adapted from BioRad literature)

BCA assay

Both the bicinchoninic acid (BCA) assay, as well as a more time-consuming assay termed the "Lowry assay", are colorimetric assays based on the reduction of Cu^{2+} to Cu^{1+} by amides. BCA is a highly selective and sensitive detection agent for Cu^{1+}. The macromolecular structure of the protein, the peptide bond number, and the presence of cysteine, cystine, tryptophan, and tyrosine contribute to the development of color in the BCA assay.

The BCA is subject to interference from some compounds, particularly reducing agents and copper chelators.

Other assays

Miscellaneous other assays are also used in assessing protein concentration. The Biuret assay is approximately 100-fold less sensitive than the Bradford assay (a more sensitive Biuret assay protocol was reported by Matsushita *et al.*[1]). Pyrogallol red-molybdate is infrequently used, with a reported sensitivity of 5-50 µg/10 µL. The "Nanoorange" reagent is a recently introduced, highly sensitive protein detection reagent.

The standard curve

---

[1]Matsushita M, T Irino, T Komoda, and Y Sakagishi, 1993, Determination of proteins by a reverse biuret method combined with the copper-bathocuproine chelate reaction, Clin Chim Acta 216, 103.

A standard curve describes the response of an assay to introduction of varying amounts of a standard, a compound representative of the sample to be assayed. The resulting data can be represented as a two-dimensional X-Y plot with the X axis representing the response and the Y axis representing the value of the standard. For many assays, the relationship between the response of the assay and the amount of standard is linear over a particular range. This is frequently considered the useful, or "working", range of the assay. A linear regression can be conducted to mathematically define the linear portion of the curve. Clearly, the more data points (in the linear range) included in the regression, the higher the confidence the investigator has in the final regression formula. With more sophisticated computing resources, all data points can be included and the appropriate non-linear regression (if known) used. Regardless, the resulting formula, $y=f(x)$, can be used to evaluate sample data. Introducing the assessed response value of a sample, x_{sample}, into the formula, yields the corresponding y value, typically representing the extrapolated concentration or mass of the sample.

Although a linear regression can be calculated by hand by the statistically competent, many calculators have statistical functions including linear regression which make the process much more rapid. It may be worth checking for these functions on your calculator or in your calculator manual. Computer programs are freely available (e.g., gnuplot) which have the capability to do relatively sophisticated regression analyses. Alternatively, a variety of online sites provide tools to conduct linear regression.

Appendix H. Spectrophotometer use

Genesys 20 Vis spectrophotometer

The wavelength range of the Genesys 20 Vis spectrophotometer is 325 - 1100 nm.

1. Turn the instrument on (the switch is located on the rear LHS). The machine should perform power-on sequence (note: ensure that cell holder is empty during the power-on sequence).

2. For a full warm-up, allow at least 30 min before use.

3. Perform Abs/Trans measurements by following the steps below:

    press [A/T/C] to select Abs or Trans mode

    press [nm up/down] to select wavelength

    insert blank into cell holder and close the sample door[1]

    press [0 ABS/100%T] to set the blank zero

    remove blank; insert sample and close the sample door

    measurement appears on the display

Figure H.1. The sample compartment and control panel of a spectrophotometer

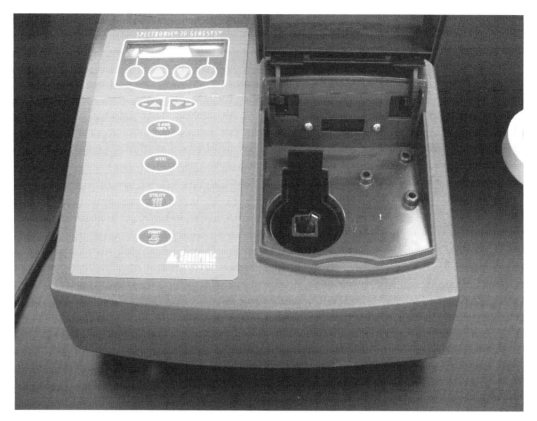

While the specifics of spectrophotometer use vary from instrument to instrument (e.g., the steps used with the Gensys 20 Vis are below), the basics are pretty similar from instrument to instrument:

- turn the instrument on, allowing time to complete any onboard diagnostics and time for the lamp(s) to "warm up"

- select the desired wavelength

Figure H.2. A cuvet

- zero the instrument on a "blank":

  - fill a cuvet (Figure H.2, "A cuvet") with the solvent in which the sample is dissolved (fill to a position at least 0.5 cm above the position corresponding to the light path)

  - place the cuvet in the spectrophotometer cuvet holder, ensuring that the transparent sides of the cuvet are situated so that they are perpendicular to the light path (i.e., so that the light path passes through the transparent faces) (note the arrow in Figure H.3, "The sample compartment of a spectrophotometer")

  - press the 'zero' or 'clear' button on the instrument (if no such control exists, record the absorbance value -- this is the value which will be subtracted from all other measurements)

Figure H.3. The sample compartment of a spectrophotometer

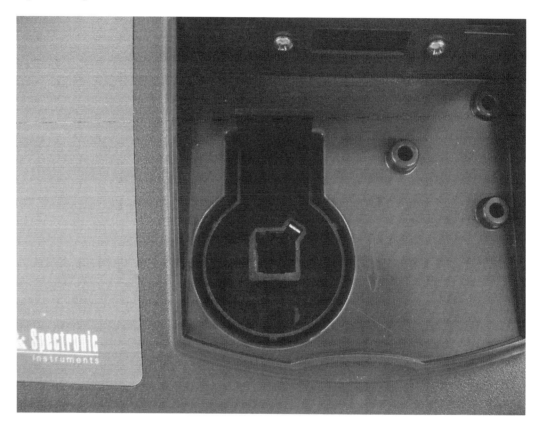

- measure the absorbance of the sample, preparing and placing the cuvet in the same manner as for the "blank"

the Genesys 20 Vis spectrophotometer

note(s): the wavelength range of the Gensys 20 Vis is 325 - 1100 nm

1. Turn the instrument on (the switch is located on the rear left-hand side). The machine should perform a power-on sequence.

Spectrophotometer use

  note: ensure that cell holder is empty during power-on sequence

2. For full warm-up, allow at least 30 min before use

3. Abs/Trans measurements:

   press [A/T/C] to select Abs or Trans mode

   press [nm up/down] to select wavelength

   insert blank into cell holder; close sample door

   note: position the cell so light passes through the clear walls of the cell

   press [0 ABS/100%T] to set the blank zero

   remove blank; insert sample

   measurement appears on the display

Appendix I. Densitometry

Several software packages are available for analyzing images such as those of gels. These include tnimage [http://brneurosci.org/tnimage.html], a Linux densitometry package, NIH Image [http://rsb.info.nih.gov/nih-image/], an image processing program for the Macintosh, ImageJ [http://rsb.info.nih.gov/ij/], a Java-based program somewhat similar to NIH Image, and Scion Image [http://www.scioncorp.com/], a Windows version of NIH Image.

Using ImageJ

Start ImageJ (in Linux, 'java -jar ij.jar')

Open the image: File → Open to import a TIFF or JPEG image

Use the "Magnifying glass" tool to zoom in (right-click (or alt-click) to zoom out)

Use Image → Rotate to rotate the image so that the lanes run vertically

If desired, use Image → Crop to crop the image

To measure distances (e.g., to calculate Rf values), use the "Line tool" to draw a line between the points of interest and use the Analyze → Measure tool to determine the length between the two points

For densitometry measurements,

- use a selection tool (e.g., the rectangle tool) to select a region of interest

Densitometry

- determine the mean grey value: Analyze → Measure (if the mean value isn't given use Analyze → Set measurements to specify which measurements are recorded)

- remember to determine the background mean grey value -- subtract it from other values to obtain the values attributable to the protein band(s)

Using Scion Image

1. Open the image: click the File pull-down and select Import to import a TIFF image

2. Use the "Magnifying glass" tool to adjust the view

3. Calibrate using loading standards and analyze sample bands of interest.

    a. Analyze → Reset

    b. Use a selection tool to record the mean gray value of each standard. Starting with the lowest standard, select the band. Click Analyze and select Measure (Ctrl - 1). Repeat with remaining standards, progressing to the highest standard.

    c. Click Analyze and select Calibrate (use a straight-line fit)

    d. Use the selection tool to record the mean gray value of each sample band of interest. Click Analyze and select Calibrate to obtain the mean gray value.

Made in the USA
Lexington, KY
12 August 2013